Recent Trends in Computational Intelligence

Edited by Ali Sadollah
and Tilendra Shishir Sinha

Published in London, United Kingdom

IntechOpen

Supporting open minds since 2005

Recent Trends in Computational Intelligence
http://dx.doi.org/10.5772/intechopen.78839
Edited by Ali Sadollah and Tilendra Shishir Sinha

Contributors
Mohamed Abdelsabour Fahmy, Ahmed Riadh BABA ALI, Aouatef Safo Mahani, Qasem AlRadaideh, Mirjam Sepesy Maučec, Gregor Donaj, Junmei Zhong, Damjan Vlaj, Marko Kos, Andrej Žgank, Ujwalla Gawande, Kamal Hajari, Yogesh Golhar, Leonid Markin

First published in London, United Kingdom, 2020 by IntechOpen
IntechOpen is the global imprint of INTECHOPEN LIMITED, registered in England and Wales, registration number: 11086078, 7th floor, 10 Lower Thames Street, London, EC3R 6AF, United Kingdom
Printed in Croatia

British Library Cataloguing-in-Publication Data
A catalogue record for this book is available from the British Library

Additional hard and PDF copies can be obtained from orders@intechopen.com

Recent Trends in Computational Intelligence
Edited by Ali Sadollah and Tilendra Shishir Sinha
p. cm.
Print ISBN 978-1-83880-705-4
Online ISBN 978-1-83880-706-1
eBook (PDF) ISBN 978-1-83880-707-8

For EU product safety concerns:
IN TECH d.o.o., Prolaz Marije Krucifikse Kozulić 3, 51000 Rijeka, Croatia, info@intechopen.com or visit our website at intechopen.com.

Meet the editors

Ali Sadollah received his BSc degree in mechanical engineering, solid states, from Azad University, Semnan Branch, Iran in 2007 and MSc degree in mechanical engineering, applied mechanics, from the University of Semnan, Semnan, Iran in 2010. He obtained his Ph.D. at the Faculty of Engineering, the University of Malaya, Kuala Lumpur, Malaysia in 2013. He has also served as a postdoctoral research fellow for more than 2 years at Korea University, Seoul, South Korea. In 2016, for a year, he served as a research staff member at Nanyang Technological University, Singapore. Afterward, he was a guest assistant professor at Sharif University of Technology, Tehran, Iran for 2 years. Currently, he is an assistant professor in the Department of Mechanical Engineering at University of Science and Culture, Tehran, Iran. His research interests include algorithm development, optimization and metaheuristics, applications of soft computing methods in engineering, artificial neural networks, and computational solid mechanics.

Dr. Tilendra Shishir Sinha, born in July 1968 in Dhanbad, Jharkhand, India, received his BE in Computer Technology from Nagpur University, Maharashtra, India, in 1992. He completed his MSc(Engineering) in Electrical Engineering (Power Electronics Specialization) from the National Institute of Technology, Jamshedpur, Jharkhand, India, in 2002. He moved on to complete his PhD(Engineering) in Computer Science & Engineering from the National Institute of Technology, Durgapur, West Bengal, India, in 2009. He has more than 25 years of experience in teaching, industry, and research. He is an IEEE member, ISTE member, and fellow member of IEI and Chartered Engineer in CSE. He has co-authored numerous publications in journals and conferences, and also published four book chapters in the area of natural language processing, speech processing, signal and image processing, information security, and cyber security. His research interests include natural language processing, promoting global cyber security, information security through biometrical patterns, underwater objects detection and recognition, digital signal and morphological image processing, and applied engineering mathematics in various applications of core engineering.

Contents

Preface

Computational intelligence (CI) is the philosophy, architecture, execution, and creation of cognitive paradigms that are biologically and linguistically driven. Neural networks, fuzzy systems, evolutionary computation, learning theory, and probabilistic methods are historically the five main pillars of CI. In recent years, applications and utilization of CI approaches have increased drastically. Applications range from natural language processing, boundary element modeling, imbalanced datasets, text mining, patient disease, human action videos, machine translation, and pedestrian detection.

In this book, CI and the applicable nature of CI are highlighted by explaining different practical applications. This book starts with some applications of CI, and then proceeds with the usage of data mining in the field of CI. The book ends with some real-life and practical applications of deep learning in various fields of studies from video detection to patient disease.

I would like to thank all the authors and scholars for their precious contributions to this book. Special thanks go to IntechOpen and Author Service Manager Mrs. Mia Vulovic for their support and patience.

Ali Sadollah, PhD
Department of Mechanical Engineering,
University of Science and Culture,
Tehran, Iran

Tilendra Shishir Sinha
Koneru Lakshmaiah Education Foundation,
India

Section 1

Computational Intelligence: Recent Trends

Chapter 1

Boundary Element Modeling and Optimization of Three Temperature Nonlinear Fractional Generalized Photo-Thermoelastic Interaction in Anisotropic Semiconductor Structures

Mohamed Abdelsabour Fahmy

Abstract

The main objective of this paper is to introduce a new fractional-order theory called nonlinear fractional generalized photo-thermoelasticity involving three temperatures. Due to strong nonlinearity, it is very difficult to solve the wave problems related to this theory analytically. Therefore, we propose a new boundary element algorithm and technique for simulation and optimization of the considered problems related to this theory. The genetic algorithm (GA) as an optimization method has been applied based on free form deformation (FFD) technique to improve the performance of our proposed technique. In the formulation of the considered problem, the profiles of the considered objects are determined by FFD technique, where the FFD control point positions are treated as genes, and then the chromosome profiles are defined with the gene sequence. The population is established by a number of individuals (chromosomes), where the objective functions of individuals are achieved by the boundary element method (BEM). A nonuniform rational B-spline curve (NURBS) was used to model optimized boundary where it reduces the number of control points and provides the flexibility to design several different shapes for solving the considered photo-thermoelastic wave problems. The numerical results verify the validity and accuracy of our proposed boundary element technique.

Keywords: boundary element method, fractional-order, nonlinear generalized photo-thermoelasticity, three temperatures, modeling and optimization, anisotropic semiconductor structures

1. Introduction

In semiconductors, an electronic deformation leads to local strain which produces plasma waves that are similar to thermal waves generated by local periodic elastic deformation. In general, the electric resistance of semiconductor decreases with increasing temperature, due to semiconductor electrons released from atoms by heat. Recently, the fractional differential equations that can be used for describing many real-world systems have gotten more and more researchers' attention due to their many applications in sciences and engineering fields.

Recently, increasing attention has been directed toward generalized micropolar thermoelastic problems in anisotropic media due to its many applications in aeronautics, astronautics, geophysics, plasma physics, nuclear plants, nuclear reactors, automobile industries, military technologies, robotics, earthquake engineering, soil dynamics, mining engineering, high-energy particle accelerators, and other engineering industries.

The classical thermoelasticity (CTE) theory has been proposed by Duhamel [1] and Neuman [2] and has two physical paradoxes. First, the heat conduction equation of this theory does not include any elastic terms. Second, the heat conduction equation is of a parabolic type, predicting infinite propagation speed of thermal energy. This prediction is a physically unacceptable situation. Biot [3] developed the classical coupled thermoelasticity (CCTE) theory to resolve the first *paradox* of CTE theory. However, both theories share the second *paradox*. So, several generalizations of Fourier's law that predicts finite propagation speed of thermal waves have been successfully developed and implemented. Lord and Shulman (L-S) [4] proposed the extended thermoelasticity (ETE) theory, where the Fourier's heat conduction law is replaced by the so-called Maxwell-Cattaneo law with one relaxation time. Green and Lindsay (G-L) [5] proposed the temperature rate-dependent thermoelasticity (TRDTE) theory including two relaxation times. Green and Naghdi (G-N) [6, 7] have formulated three different theories in the context of linear generalized thermoelasticity; the general *constitutive assumptions* for the heat flux vector in each theory are different. So, they got three models labeled as types I, II, and III. Type I is based on the classical Fourier's law of heat conduction, type II characterizes the thermoelastic behavior without energy dissipation (TEWOED), and type III describes the thermoelastic interaction with energy dissipation (TEWED). Due to the mathematical difficulties, inherent in solving coupled magnetomechanical problems [8, 9], the problems become too complicated to obtain an analytical solution in a general case. Instead of analytical methods, several numerical methods have recently been successfully developed and implemented to obtain the approximate solutions for such problems including the finite difference method (FDM) [10] and finite element method (FEM) [11]. Nowadays, the boundary element method (BEM) is an effective computational technique [12–31] which provides an excellent alternative to the prevailing finite difference and finite element methods for solving various engineering, scientific, and mathematical applications due to its simplicity, efficiency, and ease of implementation. Throughout the present paper, the new term three-temperature is presented for the first time in the field of photo-thermoelasticity.

The main aim of this paper is to introduce a new fractional-order theory called nonlinear generalized photo-thermoelasticity involving three temperatures. The governing equations of transient thermal stress wave propagation problems associated with this theory are very difficult to solve analytically because of strong nonlinearity. So, we need to develop new numerical techniques for solving such equations. Therefore, we propose a new boundary element technique for solving the governing equations of the proposed theory. The numerical results are depicted graphically to confirm the validity and accuracy of our proposed technique.

A brief summary of this chapter is as follows. Section 1 outlines the background and provides the readers with the necessary information from books and articles for a better understanding of the generalized thermoelastic theories associated with the distributions of three temperature and thermal stress fields. Section 2 describes the formulation of the new theory and its related problems. Section 3 discusses the implementation of the new BEM to obtain the carrier density field. Section 4 studies the implementation of the new BEM for solving the nonlinear radiative heat conduction equation, to obtain the three temperature fields. Section 5 studies the development of the new BEM and its implementation for solving the move equation based on the known three temperature fields, to obtain the displacement field. Section 6 discusses the shape optimization scheme for semiconductor structures. Section 7 presents the new numerical results that describe the BEM results which are in excellent agreement with the FDM and FEM results.

2. Formulation of the problem

We considered the Cartesian coordinates for a semiconductor structure which occupies the region R and bounded by a closed surface S.

The coupled plasma and thermoelastic wave equations during photothermal process can be written as follows:

The wave equation:

$$\sigma_{ij,j} + \rho F_i = \rho \ddot{u}_i \tag{1}$$

The plasma wave equation:

$$\frac{\partial N}{\partial \tau} - D_0 \nabla^2 N + \frac{1}{\tau_0}(N - n_0) = \breve{\alpha}\theta \tag{2}$$

where D_0, N, n_0, τ_0, and $\breve{\alpha}$ are the diffusion coefficient, carrier density, equilibrium carrier concentration at temperature θ, electron relaxation time, and thermal expansion coefficient, respectively. Also, we assumed that $\breve{\alpha} = \tilde{A}e^{-ax}$.

The two-dimensional three-temperature (2D-3T) radiative heat conduction equations can be expressed as follows:

$$D_\tau^a T_\alpha(r,\tau) = \xi \nabla[\mathbb{K}_\alpha \nabla T_\alpha(r,\tau)] + \xi \overline{\mathbb{W}}(r,\tau), \xi = \frac{1}{c_\alpha \rho \delta_1} \tag{3}$$

where

$$\sigma_{ij} = C_{ijkl}\delta_{ij} - \beta_{ij}\left(\theta + \frac{d_n N}{\breve{\alpha}}\right), C_{ijkl} = C_{klij} = C_{jikl}, \beta_{ij} = \beta_{ji} \tag{4}$$

$$\overline{\mathbb{W}}(r,\tau) = \begin{cases} \rho\, \mathbb{W}_{ei}(T_e - T_i) + \rho\, \mathbb{W}_{er}\left(T_e - T_p\right) + \overline{\overline{\mathbb{W}}}, \alpha = e, \delta_1 = 1 \\ -\rho\, \mathbb{W}_{ei}(T_e - T_i) + \overline{\overline{\mathbb{W}}},\ \alpha = i, \delta_1 = 1 \\ -\rho\, \mathbb{W}_{er}\left(T_e - T_p\right) + \overline{\overline{\mathbb{W}}},\ \alpha = p, \delta_1 = \dfrac{4}{\rho}T_p^3 \end{cases} \tag{5}$$

$$\begin{aligned} \overline{\overline{\mathbb{W}}}(r,\tau) = & -\delta_{2n}\mathbb{K}_\alpha \dot{T}_{\alpha,ij} + \beta_{ij}T_{\alpha 0}\left[\mathring{A}\delta_{1n}\dot{u}_{i,j} + (\tau_0 + \delta_{2n})\ddot{u}_{i,j}\right] \\ & + \rho c_\alpha\left[(\tau_0 + \delta_{1n}\tau_2 + \delta_{2n})\ddot{T}_\alpha\right] - \frac{E_g}{\tau_0}(N - n_0) \end{aligned} \tag{6}$$

where

$$\mathbb{W}_{ei} = \rho\mathbb{A}_{ei}T_e^{-2/3}, \mathbb{W}_{er} = \rho\mathbb{A}_{er}T_e^{-1/2}, \mathbb{K}_\alpha = \mathbb{A}_\alpha T_\alpha^{5/2}, \alpha = e, i, \mathbb{K}_p = \mathbb{A}_p T_p^{3+\mathbb{B}} \tag{7}$$

The total energy of unit mass can be described by

$$P = P_e + P_i + P_p, P_e = c_e T_e, P_i = c_i T_i, P_p = \frac{1}{\rho} c_p T_p^4 \tag{8}$$

where σ_{ij} is mechanical stress tensor; ρ is the density; F_i is the mass force vector; u_i is the displacement vector; C_{ijkl} is the constant elastic moduli; β_{ij} are the stress-temperature coefficients; c_e, c_i, and c_p are specific heat capacities of electron, ion, and phonon, respectively; $\mathbb{K}_e, \mathbb{K}_i$, and $\mathbb{K}_p$ are conductive coefficients of electron, ion, and phonon, respectively; $\mathbb{W}_{ei}$ is the electron-ion coefficient; $\mathbb{W}_{ep}$ is the electron-phonon coefficient; the total temperature $\theta = T_e + T_i + T_p$, $-\frac{E_g}{\tau_0}(N - n_0)$ is the recombination term; and E_g is the semiconductor gap energy.

3. BEM solution of carrier density field

In order to construct the integral equation, we use the following Green's function:

$$G(x, \tau) = \frac{e^{-\frac{\tau}{\tau_0}}}{2\sqrt{(\pi D_0 \tau)}} e^{-x^2/4D_0\tau}. \tag{9}$$

We assume that the solution of Eq. (2) can be written as

$$N = n_0 + N'(x, \tau) + \int g(\tau')G(x, \tau - \tau')d\tau', \tag{10}$$

where $G(x, \tau - \tau')$ is a particular solution of Eq. (2) when its right-hand side is equal to zero and $N'(x, \tau)$ is also a particular solution of Eq. (2) which can be obtained as

$$N'(x, \tau) = A\int_0^\tau d\tau' \int_{-\infty}^{\infty} e^{-ax} G(x - x', \tau - \tau')dx'. \tag{11}$$

which can be written in the following form [32].

$$N'(x, \tau) = Ae^{-ax} \frac{\tau_0}{1 - a^2 L_0^2}\left[1 - e^{-\left(1 - a^2 L_0^2\right)\tau/\tau_0}\right], \tag{12}$$

where the minority carrier diffusion length is $L_0 = \sqrt{D_0 \tau_0}$. Thus after imposing initial conditions ($N(x, 0) = N_0$ for all x) and boundary conditions ($N(0, \tau) = N_0$ for all τ), we have

$$\int_0^\tau g(\tau')G(0, \tau - \tau')d\tau' = -N'(0, \tau) \tag{13}$$

By solving Eq. (13), the unknown $g(\tau)$ is determined. Then from Eq. (10), we obtain $N(x, \tau)$.

4. BEM solution of temperature field

By applying the Caputo scheme, we have [33].

$$D_\tau^a T_\alpha^{f+1} + D_\tau^a T_\alpha^f \approx \sum_{j=0}^{k} W_{a,j}\left(T_\alpha^{f+1-j}(r) - T_\alpha^{f-j}(r)\right), (\mathrm{f} = 1, 2, \dots, \mathrm{F}), \tag{14}$$

where

$$W_{a,0} = \frac{(\Delta\tau)^{-a}}{\Gamma(2-a)}, W_{a,j} = W_{a,0}\left((j+1)^{1-a} - (j-1)^{1-a}\right), j = 1, 2, \dots, F.$$

Substituting Eq. (11) into Eq. (3), we obtain

$$\begin{aligned} &W_{a,0} T_\alpha^{f+1}(r) - \mathbb{K}_\alpha(x) T_{\alpha,II}^{f+1}(r) - \mathbb{K}_{\alpha,I}(x) T_{\alpha,I}^{f+1}(r) \\ &= W_{a,0} T_\alpha^f(r) - \mathbb{K}_\alpha(x) T_{\alpha,II}^f(r) - \mathbb{K}_{\alpha,I}(x) T_{\alpha,J}^f(r) \\ &- \sum_{j=1}^{f} W_{a,j}\left(T_\alpha^{f+1-j}(r) - T_\alpha^{f-j}(r)\right) + \overline{\mathbb{W}}_m^{f+1}(x,\tau) + \overline{\mathbb{W}}_m^f(x,\tau), f = 0, 1, 2, \dots, F. \end{aligned} \tag{15}$$

Based on the fundamental solution which satisfies Eq. (15), the boundary integral equations corresponding to Eq. (3) can be expressed as

$$CT_\alpha = \int_S \left[T_\alpha q^* - T_\alpha^* q\right] dS - \int_R \frac{\mathbb{K}_\alpha}{D} \frac{\partial T_\alpha^*}{\partial \tau} T_\alpha \, dR. \tag{16}$$

Based on [34], we can write

$$C\,\dot{T}_\alpha + H\,T_\alpha = G\,Q \tag{17}$$

To solve Eq. (17), the functions T_α and q can be interpolated as

$$T_\alpha = (1-\theta) T_\alpha^m + \theta\, T_\alpha^{m+1}, \tag{18}$$

$$q = (1-\theta) q^m + \theta\, q^{m+1}. \tag{19}$$

Differentiating Eq. (18) with time, we obtain

$$\dot{T}_\alpha = \frac{dT_\alpha}{d\theta}\frac{d\theta}{d\tau} = \frac{T_\alpha^{m+1} - T_\alpha^m}{\tau^{m+1} - \tau^m} = \frac{T_\alpha^{m+1} - T_\alpha^m}{\Delta\tau^m}, \theta = \frac{\tau - \tau^m}{\tau^{m+1} - \tau^m}, 0 \le \theta \le 1. \tag{20}$$

By substituting Eqs. (18)-(20) into Eq. (17), we get

$$\left(\frac{C}{\Delta\tau^m} + \theta H\right) T_\alpha^{m+1} - \theta G Q^{m+1} = \left(\frac{C}{\Delta\tau^m} - (1-\theta) H\right) T_\alpha^m + (1-\theta) G Q^m. \tag{21}$$

Thus, the temperature can be determined from the following system:

$$\mathbb{a}\mathrm{X} = \mathbb{b}, \tag{22}$$

where $\mathbb{a}$ is an unknown matrix and X and $\mathbb{b}$ are known matrices.

5. BEM solution of displacement field

On the basis of the weighted residual method, the differential equations (1) can be transformed to the following integral equations:

$$\int_R (\sigma_{ij,j} + U_i)u_i^* \, dR = 0, i,j = 1, 2, \ldots, N \tag{23}$$

in which

$$U_i = \rho F_i - \rho \ddot{u}_i. \tag{24}$$

According to Huang and Liang [35], Eringen [36], and Dragos [37], we can write Eq. (23) as

$$C^n \mathbb{q}^n = \sum_{j=1}^{N_e} \left[-\int_{\Gamma_j} \mathbb{p}^* \psi \, d\Gamma \right] \mathbb{q}^j + \sum_{j=1}^{N_e} \left[\int_{\Gamma_j} \mathbb{q}^* \psi \, d\Gamma \right] \mathbb{p}^j, \mathbb{q} = \psi \, \mathbb{q}^j, \mathbb{p} = \psi \, \mathbb{p}^j \tag{25}$$

which can be written as

$$C^i \mathbb{q}^i = -\sum_{j=1}^{N_e} \hat{\mathbb{H}}^{ij} \mathbb{q}^j + \sum_{j=1}^{N_e} \hat{\mathbb{G}}^{ij} \mathbb{p}^j. \tag{26}$$

This matrix system can be written as follows:

$$\mathbb{H}\mathbb{Q} = \mathbb{G}\mathbb{P}, \tag{27}$$

where $\mathbb{Q}$ represents the displacements and $\mathbb{P}$ represents the tractions.
By using the boundary conditions in Eq. (27), we get

$$\mathbb{A}\,\mathbb{X} = \mathbb{B}, \tag{28}$$

where $\mathbb{A}$ is an unknown matrix and $\mathbb{X}$ and $\mathbb{B}$ are known matrices. We refer the interested readers to Reference [37] for further details.

6. Shape optimization scheme for semiconductor structures

Two criteria can be implemented during shape optimization of semiconductor structures:

I. The minimum global compliance based on the tractions λ and boundary displacements u

$$\mathcal{F} = \frac{1}{2}\int_S (\lambda \bullet u) \, dS, \tag{29}$$

II. The minimum boundary based on the equivalent stresses σ_{ij} and the reference stress σ_0

$$\mathcal{F} = \int_S \left(\frac{\sigma_{ij}}{\sigma_0}\right)^n dS, \tag{30}$$

where n is a natural number.

Based on the boundary displacement u and the reference displacement u_0, we can write

$$\mathcal{F} = \int_S \left(\frac{u}{u_0}\right)^n dS, \tag{31}$$

which can be used to obtain

$$\mathcal{F} = \delta \sum_{k=1}^{M} \left(u^k - \hat{u}^k\right) + \eta \sum_{l=1}^{N} \left(\theta^l - \hat{\theta}^l\right). \tag{32}$$

The efficiency of the proposed technique has been improved using FFD, GA, and the following nonuniform rational B-spline curve (NURBS):

$$C(t) = \frac{\sum_{i=0}^{n} N_{i,o}(t)\varpi_i P_i}{\sum_{i=0}^{n} N_{i,o}(t)\varpi_i}, \tag{33}$$

where $N_{i,o}(t)$ are the B-spline basis functions of order o and ϖ_i are the weights of control points P_i.

7. Numerical results and discussion

The efficiency of our numerical modeling technique has been improved using a nonuniform rational B-spline curve (NURBS) to decrease the computation time and the model's optimized boundary where it reduces the number of control points and provides the flexibility to design a large variety of shapes.

Figure 1 shows the main steps of the genetic algorithm of photo-thermoelastic semiconductor structures.

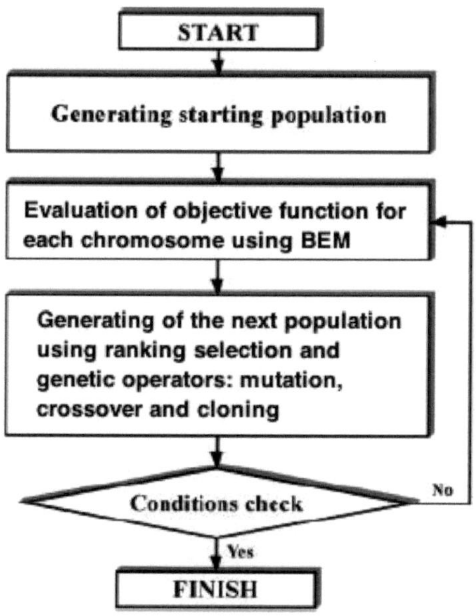

Figure 1.
Genetic algorithm of photo-thermoelastic semiconductor structures.

The design vector is represented by a chromosome x which consists of genes x_i, i = 1, ... , N:

$$x = [x_1, \ldots, x_i, \ldots, x_N] \tag{34}$$

Thus, genes can be considered as design variables.

The following constraints are also imposed on each gene:

$$x_{iL} \leq x_i \leq x_{iR}, i = 1, \ldots, N \tag{35}$$

where x_{iL} and x_{iR} are the left and right admissible values of x_i.

The uniform mutation and boundary mutation are implemented, where the uniform mutation operator replaces a gene of the chromosome with the new random value x_i which corresponds to the design parameter as shown in **Figure 2**.

The uniform mutation probability determines the gene number which will be modified in each population. The boundary mutation operator is a special case of the uniform mutation. The gene after mutation receives one of the boundary values x_{iL} or x_{iR} as shown in **Figure 3**.

The boundary mutation is very useful for boundary element problems in which the solution is on the boundary. The boundary mutation probability determines the gene number which will be modified in each population.

The simple crossover and arithmetical crossover are implemented, where the operator of the simple crossover creates two new chromosomes x′ and y′ from two existing chromosomes selected randomly, x and y, where both chromosomes are coupled together as shown in **Figure 4**.

The simple crossover probability determines the chromosomes number which will be crossing in each population.

The arithmetic crossover operator creates two identical new chromosomes x′ from two existing chromosomes selected randomly, x and y, where the gene values in the new chromosomes are the arithmetic average of genes of the parents as shown in **Figure 5**.

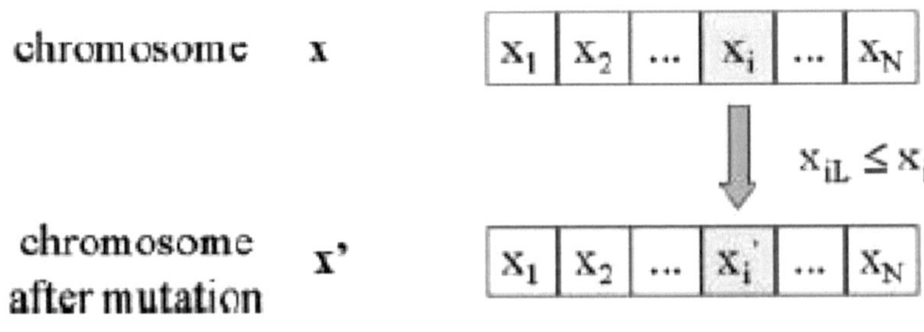

Figure 2.
Implementation of uniform mutation.

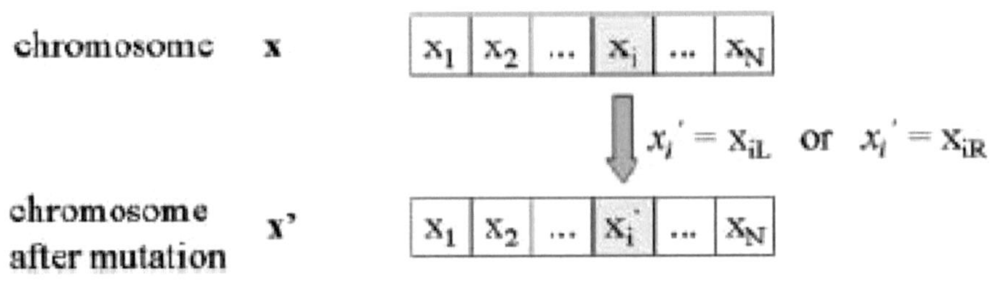

Figure 3.
Implementation of boundary mutation.

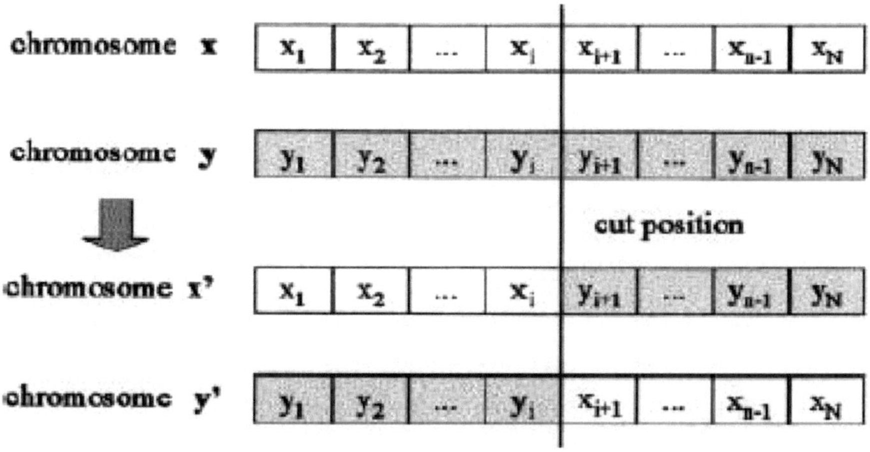

Figure 4.
Implementation of simple crossover.

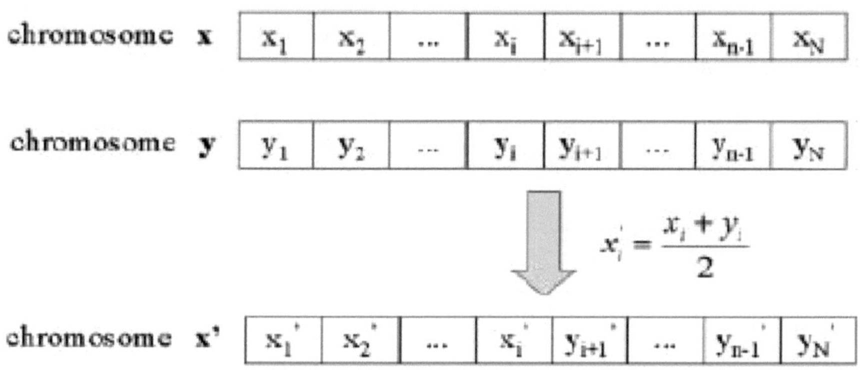

Figure 5.
Implementation of arithmetic crossover.

The operator of the cloning increases the probability of survival of the best chromosome by duplicating this one to the next generation. The probability of the cloning decides how many copies of the best chromosome will be in the new generation.

The ranking selection allows chromosomes to survive with a great value of an objective function. The first step of the ranking selection is sorting all the chromosomes according to the value of the objective function. Then on the basis of the position in the population, the probability of survival is attributed to every chromosome by the following formula:

$$\text{prob (rank)} = q(1-q)^{\text{rank}-1} \tag{36}$$

where rank is the chromosome position after sorting, prob (rank) is the probability of survival, and q is a selection coefficient.

A shape optimization of the photo-thermoelastic semiconductor structure presented in **Figure 6** is considered. Only the parts of the boundary, where the temperature field T_0 and the heat flux q_0 are prescribed, undergo the shape modification.

The optimal shape of the photo-thermoelastic semiconductor structure for isotropic, transversely isotropic, orthotropic, and anisotropic is presented in **Figure 7**. **Table 1** contains the genetic algorithm parameters which were applied.

The efficiency of our numerical modeling technique has been improved using GA, FFD, and NURBS to decrease the computation time of solving three-temperature photo-thermoelastic problems in semiconductor structures. Due to strong nonlinearity, it is very difficult to solve the problems related to this theory analytically. Therefore, we propose a new boundary element model for our current complex problem. So, the validity and accuracy of the proposed technique were confirmed by comparing graphically the one-dimensional results obtained from BEM with those obtained using the finite difference method (FDM) of Pazera and

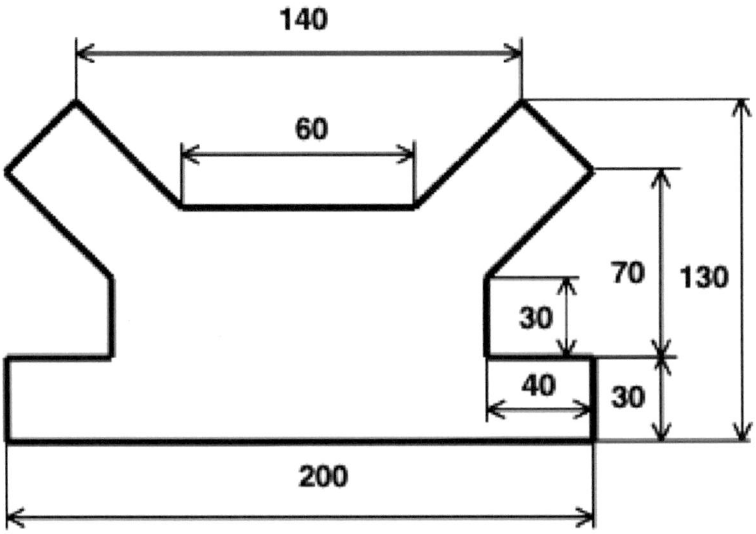

Figure 6.
Optimized considered photo-thermoelastic semiconductor structure.

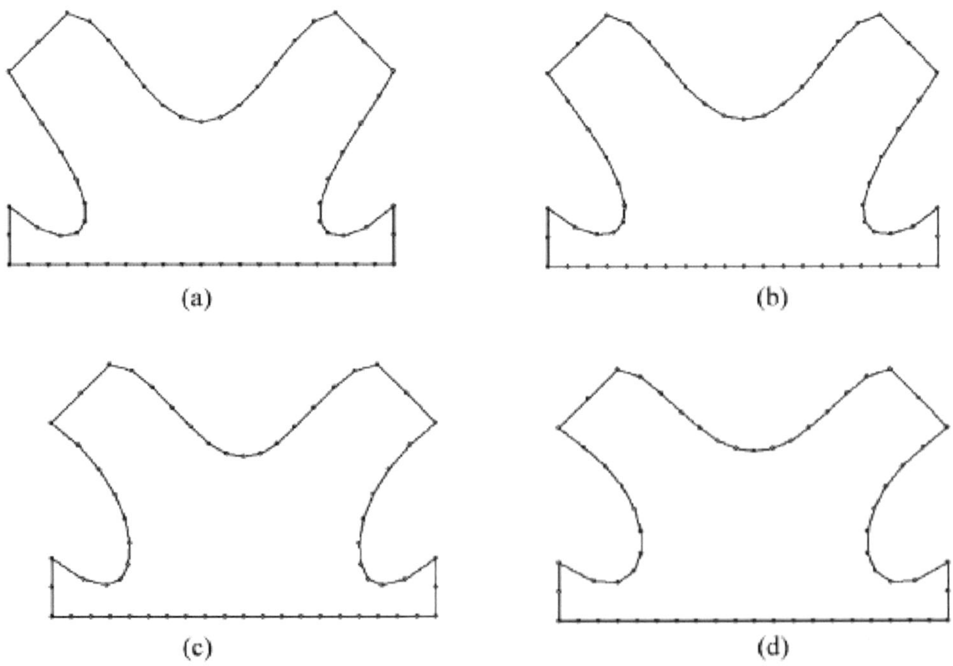

Figure 7.
Optimal shape of photo-thermoelastic semiconductor structure. (a) Isotropic, (b) transversely isotropic, (c) orthotropic, and (d) anisotropic.

Chromosome number	100
Iteration number	150
Design parameter number	5
Uniform mutation probability	0.015
Boundary mutation probability	0.0075
Simple crossover probability	0.075
Arithmetic crossover probability	0.075
Cloning probability	0.05
Selection coefficient	0.1

Table 1.
Parameters of genetic algorithm.

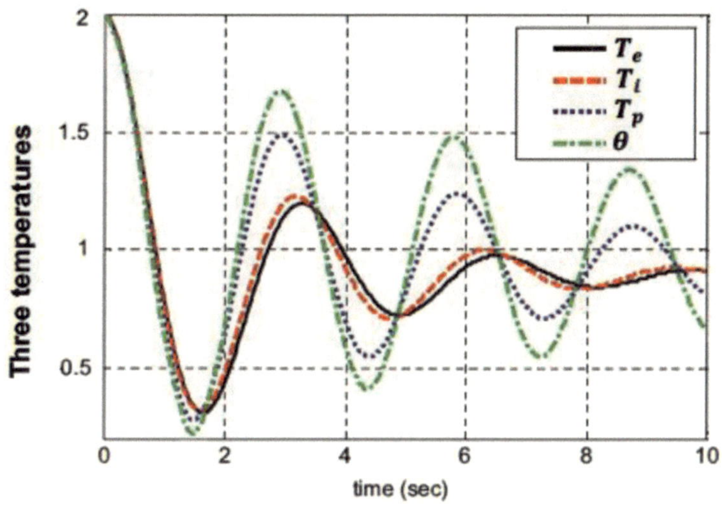

Figure 8.
Variation of the three temperatures T_e, T_i, and T_p with time τ.

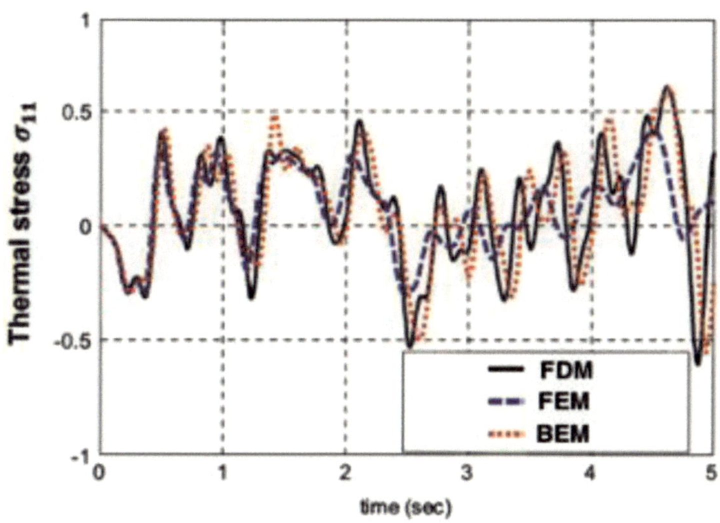

Figure 9.
Variation of the thermal stress σ_{11} with time τ.

Jędrysiak [38] and finite element method (FEM) of Xiong and Tian [39] which have been reduced as a special case from the current problem. For comparison reasons, the 2D-3T radiative heat conduction is replaced by heat conduction. **Figure 8** shows the variations of the temperature T_e, T_i, T_p and $\theta = T_e + T_i + T_p$ with the time τ. The differences between time distributions of electron temperature T_e, ion temperature T_i, phonon temperature T_p, and total temperature θ can be seen from this figure. **Figures 9–11** show the variations of the thermal stresses σ_{11}, σ_{12}, and σ_{22} with the time τ. It can be seen from these figures that the BEM results are in excellent agreement with the FDM and FEM results.

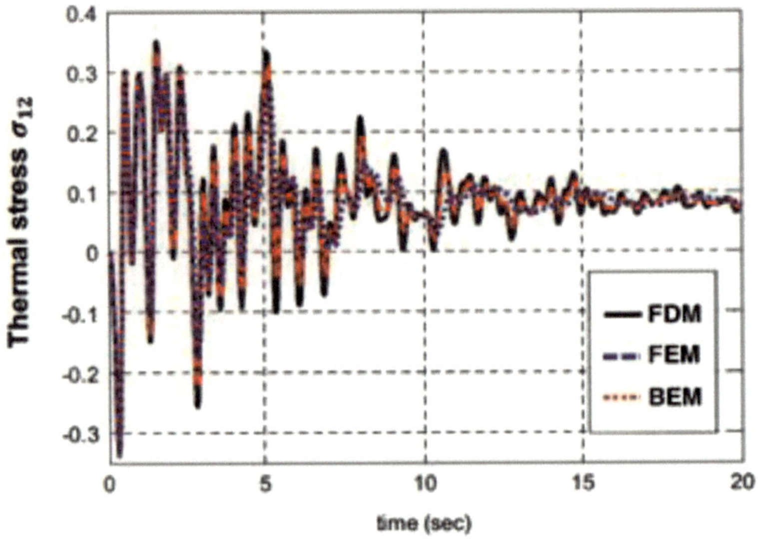

Figure 10.
Variation of the thermal stress σ_{12} with time τ.

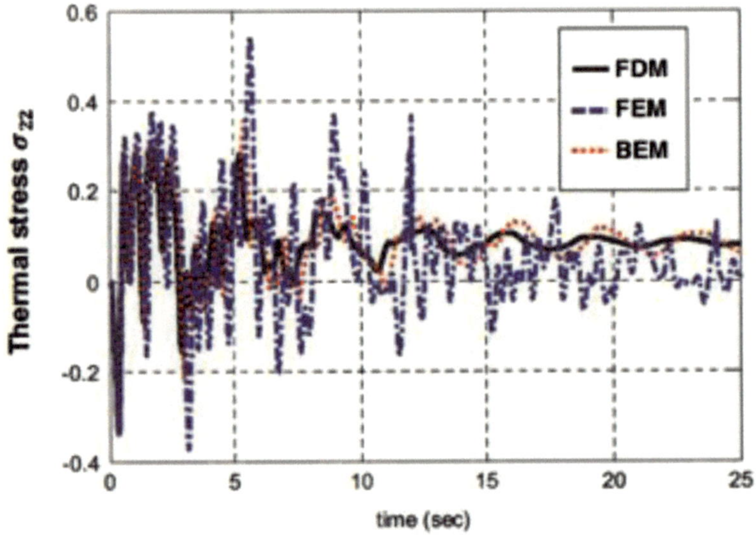

Figure 11.
Variation of the thermal stress σ_{22} with time τ.

8. Conclusion

The aim of this study is to propose a new theory called nonlinear fractional generalized photo-thermoelasticity involving three temperatures and implement a new boundary element technique for modeling and optimization of the three-temperature nonlinear fractional generalized photo-thermoelastic interaction problems in anisotropic semiconductor structures associated with the proposed theory. This technique is implemented based on genetic algorithm (GA), free-form deformation (FFD) method, and nonuniform rational B-spline curve (NURBS) as the global optimization techniques for solving complex problems associated with the proposed theory. FFD is an efficient and accurate technique for treating optimization problems with complex shapes. In the formulation of the considered problem, solutions are obtained for specific arbitrary parameters which are the control point positions in the considered problem; the profiles of the considered objects are determined by FFD method, where the FFD control points positions are treated as genes; and then the chromosomes profiles are defined with the gene sequence. The population is founded by a number of individuals (chromosomes), where the objective functions of individuals are determined by the BEM. The optimal shape of the photo-thermoelastic semiconductor structure for isotropic, transversely isotropic, orthotropic, and anisotropic is obtained. The proposed technique can be applied to a wide range of modeling and optimization problems related with our proposed theory. The numerical results verify the validity and accuracy of our proposed boundary element technique. Also, the BEM is more powerful and simple to use than the FDM or FEM, because it reduces the computational cost. The present numerical results for our general and complex problem may provide interesting information for mechanical engineers, material science researchers, computer scientists, and designers of semiconductor devices.

Author details

Mohamed Abdelsabour Fahmy
Faculty of Computers and Informatics, Suez Canal University, Ismailia, Egypt

*Address all correspondence to: mohamed_fahmy@ci.suez.edu.eg

References

[1] Duhamel J. Some memoire sur les phenomenes thermo-mechanique. Journal de l'École polytechnique. 1837;**15**:1-57

[2] Neumann F. Vorlesungen Uber die theorie der elasticitat. Brestau: Meyer; 1885

[3] Biot M. Thermoelasticity and irreversible thermo-dynamics. Journal of Applied Physics. 1956;**27**:249-253

[4] Lord HW, Shulman Y. A generalized dynamical theory of thermoelasticity. Journal of the Mechanics and Physics of Solids. 1967;**15**:299-309

[5] Green AE, Lindsay KA. Thermoelasticity. Journal of Elasticity. 1972;**2**:1-7

[6] Green AE, Naghdi PM. On undamped heat waves in an elastic solid. Journal of Thermal Stresses. 1992;**15**:253-264

[7] Green AE, Naghdi PM. Thermoelasticity without energy dissipation. Journal of Elasticity. 1993; **31**:189-208

[8] Kaliski S. Thermo-magneto-microelasticity. Bulletin of the Polish Academy of Sciences: Technical Sciences. 1968;**16**:7-12

[9] Jafarian A, Ghaderi P, Golmankhaneh AK, Baleanu D. Analytic solution for a nonlinear problem of magneto-thermoelasticity. Reports on Mathematical Physics. 2013;**71**:399-411

[10] Abd-Alla AM, El-Naggar AM, Fahmy MA. Magneto-thermoelastic problem in non-homogeneous isotropic cylinder. Heat and Mass Transfer. 2003; **39**:625-629

[11] Abbas IA, Abd-alla AN, Othman MIA. Generalized magneto-thermoelasticity in a fiber-reinforced anisotropic half-space. International Journal of Thermophysics. 2011;**32**: 1071-1085

[12] Fahmy MA. Thermoelastic stresses in a rotating non-homogeneous anisotropic body. Numerical Heat Transfer: Part A Applications. 2008;**53**:1001-1011

[13] Fahmy MA. A time-stepping DRBEM for magneto-thermo-viscoelastic interactions in a rotating nonhomogeneous anisotropic solid. International Journal of Applied Mechanics. 2011;**3**:1-24

[14] Fahmy MA. A time-stepping DRBEM for the transient magneto-thermo-visco-elastic stresses in a rotating non-homogeneous anisotropic solid. Engineering Analysis with Boundary Elements. 2012;**36**:335-345

[15] Fahmy MA. Transient magneto-thermoviscoelastic plane waves in a non-homogeneous anisotropic thick strip subjected to a moving heat source. Applied Mathematical Modelling. 2012; **36**:4565-4578

[16] Fahmy MA. Numerical modeling of transient magneto-thermo-viscoelastic waves in a rotating nonhomogeneous anisotropic solid under initial stress. International Journal of Modeling, Simulation, and Scientific Computing. 2012;**3**:1250002

[17] Fahmy MA. The effect of rotation and inhomogeneity on the transient magneto-thermoviscoelastic stresses in an anisotropic solid. Journal of Applied Mechanics. 2012;**79**:1015

[18] Fahmy MA. Transient magneto-thermo-viscoelastic stresses in a rotating nonhomogeneous anisotropic solid with and without a moving heat source. Journal of Engineering Physics and Thermophysics. 2012;**85**:950-958

[19] Fahmy MA. Transient magneto-thermo-elastic stresses in an anisotropic viscoelastic solid with and without moving heat source. Numerical Heat Transfer: Part A Applications. 2012;**61**:547-564

[20] Fahmy MA. Implicit-explicit time integration DRBEM for generalized magneto-thermoelasticity problems of rotating anisotropic viscoelastic functionally graded solids. Engineering Analysis with Boundary Elements. 2013; **37**:107-115

[21] Fahmy MA. Generalized magneto-thermo-viscoelastic problems of rotating functionally graded anisotropic plates by the dual reciprocity boundary element method. Journal of Thermal Stresses. 2013;**36**:1-20

[22] Fahmy MA. A three-dimensional generalized magneto-thermo-viscoelastic problem of a rotating functionally graded anisotropic solids with and without energy dissipation. Numerical Heat Transfer: Part A Applications. 2013;**63**:713-733

[23] Fahmy MA. A computerized DRBEM model for generalized magneto-thermo-visco-elastic stress waves in functionally graded anisotropic thin film/substrate structures. Latin American Journal of Solids and Structures. 2014;**11**:386-409

[24] Fahmy MA. Computerized Boundary Element Solutions for Thermoelastic Problems: Applications to Functionally Graded Anisotropic Structures. Saarbrücken: LAP Lambert Academic Publishing; 2017

[25] Fahmy MA. Boundary Element Computation of Shape Sensitivity and Optimization: Applications to Functionally Graded Anisotropic Structures. Saarbrücken: LAP Lambert Academic Publishing; 2017

[26] Fahmy MA. Shape design sensitivity and optimization for two-temperature generalized magneto-thermoelastic problems using time-domain DRBEM. Journal of Thermal Stresses. 2018;**41**: 119-138

[27] Fahmy MA. Shape design sensitivity and optimization of anisotropic functionally graded smart structures using bicubic B-splines DRBEM. Engineering Analysis with Boundary Elements. 2018;**87**:27-35

[28] Fahmy MA. Boundary element algorithm for modeling and simulation of dual phase lag bioheat transfer and biomechanics of anisotropic soft tissues. International Journal of Applied Mechanics. 2018;**10**:1850108

[29] Fahmy MA. Modeling and optimization of anisotropic viscoelastic porous structures using CQBEM and moving asymptotes algorithm. Arabian Journal for Science and Engineering. 2019;**44**:1671-1684

[30] Fahmy MA. Boundary element modeling and simulation of biothermomechanical behavior in anisotropic laser-induced tissue hyperthermia. Engineering Analysis with Boundary Elements. 2019;**101**:156-164

[31] Fahmy MA. A new LRBFCM-GBEM modeling algorithm for general solution of time fractional order dual phase lag bioheat transfer problems in functionally graded tissues. Numerical Heat Transfer: Part A Applications. 2019;**75**:616-626

[32] De Mey G. An integral equation method to calculate the transient behavior of a photovoltaic solar cell. Solid-State Electronics. 1978;**21**:595-596

[33] Cattaneo C. Sur une forme de i'equation de la chaleur elinant le paradox d'une propagation instantanc. Comptes Rendus de l'Académie des Sciences. 1958;**247**:431-433

[34] Wrobel LC, Brebbia CA. The dual reciprocity boundary element

formulation for nonlinear diffusion problems. Computer Methods in Applied Mechanics and Engineering. 1987;**65**:147-164

[35] Huang FY, Liang KZ. Boundary element method for micropolar thermoelasticity. Engineering Analysis with Boundary Elements. 1996;**17**:19-26

[36] Eringen AC. Theory of micropolar elasticity. In: Liebowitz H, editor. Fracture. New York: Academic Press; 1968

[37] Dragos L. Fundamental solutions in micropolar elasticity. International Journal of Engineering Science. 1984;**22**: 265-275

[38] Pazera E, Jędrysiak J. Effect of microstructure in thermoelasticity problems of functionally graded laminates. Composite Structures. 2018; **202**:296-303

[39] Xiong QL, Tian XG. Generalized magneto-thermo-microstretch response during thermal shock. Latin American Journal of Solids and Structures. 2015; **12**:2562-2580

Chapter 2

Using Discrete Geometric Models in an Automated Layout

Leonid V. Markin

Abstract

The application of discrete (voxel) geometric models in computer-aided design problems is shown. In this case, the most difficult formalized task of computer-aided design is considered—computer-aided layout. The solution to this problem is most relevant when designing products with a high density of layout (primarily transport equipment). From a mathematical point of view, these are placement problems; therefore, their solution is based on the use of a geometric modeling apparatus. The basic provisions and features of discrete modeling of geometric objects, their place in the system of geometric modeling, the advantages and disadvantages of discrete geometric models, and their primary use are described. Their practical use in solving some of the practical problems of automated layout is shown. This is the definition of the embeddability of the placed objects and the task of tracing and evaluating the shading. Algorithms and features of their practical implementation are described. A numerical assessment of the accuracy and performance of the developed geometric modeling algorithms shows the possibility of their implementation even on modern computers of medium power. This allows us to hope for the integration of the developed layout algorithms into modern systems of solid-state geometric modeling in the form of plug-ins.

Keywords: geometric model, discreteness, layout, computer-aided design, voxel (receptor), embeddability, trace, shadowing, algorithm, heuristics, multivalued logic

1. Introduction

In the process of designing any technical structures according to literature data, 80–90% of the processed information is geometric information about the shape of the designed product. In some cases, the requirements for the accuracy of the description must be very high, for example, when modeling the flow around objects or automating technological processes (to ensure specified gaps during assembly of products or reproduction-oriented products on equipment with numerical program control). All this causes to a wide range of geometric modeling methods used in modern technology.

Theoretical questions of the geometric modeling method from various aspects were investigated in the works of Russian scientists—Valkova K.I., Ivanova G.S., Kotova I.I., Osipova V.A., Yakunina V.I., Rvacheva V.L. etc.,—and the works of Robert Fergusson, Stephen Coons, Pierre Bezier, Charles Hermite, Isaac Jacob Schoenberg, Carl de Boor, Ken Versprille, Eugene Lee, Steve Ginsberg, and others. Their works contain both classical and computer-oriented methods of assignment, calculation, and reproduction of geometric forms of the designed objects.

Of course, such an abundance of methods is focused on the description of the geometric shape of heterogeneous technical objects. To classify the geometric models used in the description of design objects, it is advisable to use the approach, proposed by Semenkov O.I.-Osipov V.A. in [1], which is based on the structure of the synthesis of geometric objects from their constituent elements. This classifier divides all geometric objects (GO) into two large groups—geometric objects of complex technical form and geometric objects of complex technical structure. Objects of the first group are limited to compartments of surfaces, each of which is described by sufficiently complex analytical equations or systems of equations. These include aircraft fuselages, car bodies, ship hulls, turbine blades, etc. The objects of the second group are combined on the basis of set-theoretic operations (union, intersection, negation) are described, as a rule, relatively simple geometric shapes. In Markin [2], we proved that based on the specifics of the solved project tasks, the number of such groups should be increased to four (**Figure 1**).

The abundance of geometric forms of objects in engineering, construction, design, etc. requires a library of geometric modeling methods adapted to describe the specific features of the shape of geometric objects. Therefore, in addition to the classification of geometric objects, there are classification systems of geometric modeling methods themselves, which can be divided into three classes (**Figure 2**).

Sculptural methods are used to create geometric models of such objects, and the exact analytical description of which is unknown and can hardly be obtained. In addition to design, sculptural methods are widely used in engineering (aviation, shipbuilding, and automotive), when the shape of the surface is corrected not only for esthetic reasons but also on the basis of aerodynamic or hydrodynamic experiment (**Figure 3a**). However, in the end we get an analytical expression of the geometric shape of these objects with varying degrees of accuracy.

The implementation of this method is based on a fairly large library of surface approximation methods using splines, B-splines, NURBS, Koon's surfaces, Hermite, Lagrange, Bezier, etc.

Analytical approximation methods are used to describe the shape of objects consisting of complex surfaces of the second and higher orders. Since the direct computational processing of surfaces of such a complex geometric shape is difficult, they are approximated by areas of surfaces of lower order (planes, cylinders, spheres, etc.) (**Figure 3b**).

Accurate methods of modeling three-dimensional objects are a set of the following known methods:

- Kinematic
- Parametric
- Wireframe
- Piecewise analytical
- Algebra-logical (R-function method)
- The method of "decomposition into elements"
- The method of constructive geometry of the elementary volume

An illustration of these methods of geometric modeling is shown in **Figure 3b** and **c**.

DOI: http://dx.doi.org/10.5772/intechopen.90941

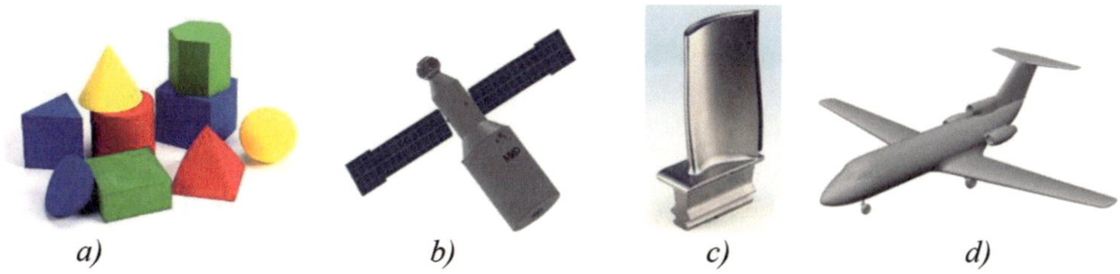

Figure 1.
Classification of geometric objects by level of complexity: (a) primitives, (b) objects of complex technical structures, (c) objects of complex technical forms, and (d) objects of complex technical form and structure.

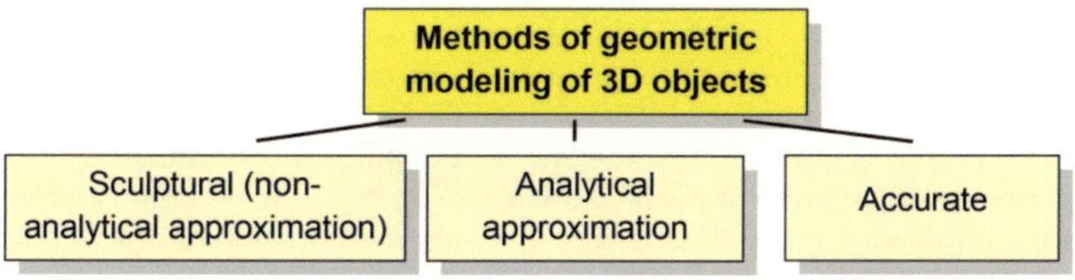

Figure 2.
Classification of methods for modeling geometric objects.

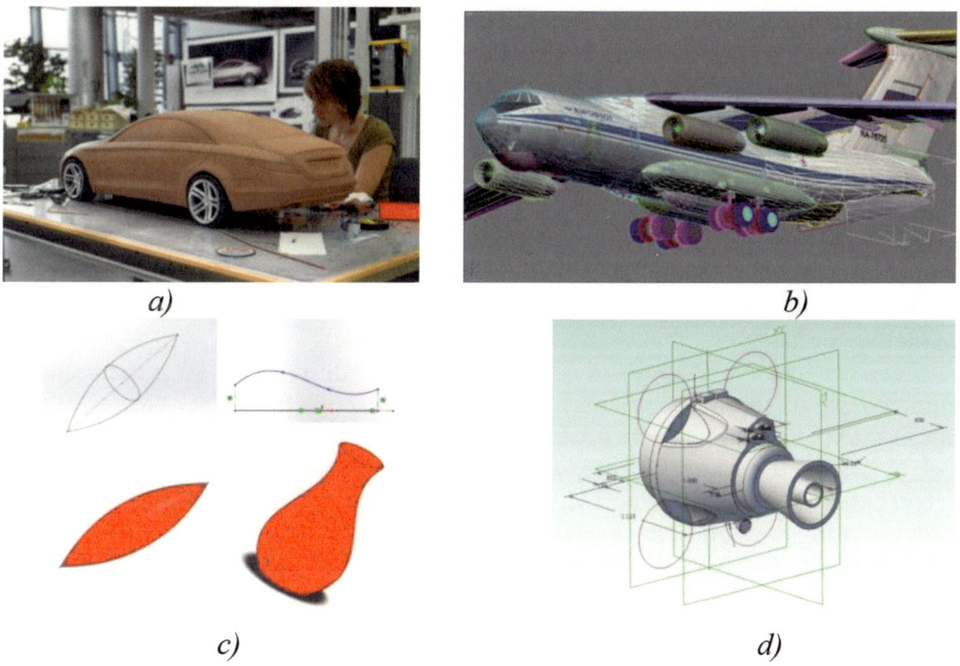

Figure 3.
Examples of technical objects implemented by sculptural methods: (a) analytical approximation methods; (b) exact methods; kinematic (c) and parametric (d).

2. Computer-aided design problems focused on discrete geometric models

Deniskin et al. [3] showed that at present the problem of computer representation of the geometric shape of objects of any complexity can be considered sufficiently solved. However, in some design tasks, the requirement of accuracy of the description of the form is not the main one. An example of such tasks is the automation of layout calculations. In modern technology, the quality of the layout (i.e., placing the necessary equipment and payload) largely determines the quality of the design.

The development of technology, primarily transport and, especially, aerospace, increasing the density of the layout makes designers constantly improve the methods of design automation. For illustration, two aircrafts of different eras are shown with approximately the same takeoff mass. (30 tons)—ANT-20 (30 years of the last century—**Figure 4a**) and a modern Su-24 (**Figure 4b**). At the same time, much more various onboard equipment is installed on a modern aircraft.

Many requirements must be complied when designing the layout (both automated and traditional methods)—ensuring maximum density, but with the exception of mutual crossing of the hosted objects—to provide a predetermined position of the center of mass of the composed object and the minimum bulk of communications between placed objects, with the exception of the proximity of incompatible objects (such as "hot at work" and "cold"). An additional requirement is to ensure the ergonomics of the layout (the possibility of installation and maintenance of equipment). It is also necessary to ensure the reliability of the functioning of the assembled equipment, which depends on the levels of vibration, pressure drop, temperature, etc., which in turn is determined by the location of this in the designed vehicle.

Taking into account so many of these factors that determine the quality of layout solutions requires either a lot of engineering experience of the developer or the use of information technology in solving this problem. For objects with a high-density layout, even the most careful placement on the drawing does not exclude the possibility of cases of their mutual intersection. To avoid this situation allows the creation of the physical layout, which in scale or life-size simulated layout of a solution. However, despite the attractiveness of physical models, their production is long and expensive. Therefore, the development of methodological, algorithmic, and software processes of automated placement is an actual practical task. Since object placement problems are geometric problems (in geometry they are called positional), their solution should be sought in an extensive library of geometric modeling methods.

Even the first experiments of computerization of design process at the decision of separate private problems have shown their high efficiency. Work on the automation of placement was no exception. The first publications in this area belong to the 60 of the last century and are associated with the names of Russian scientists, L.V. Kantorovich and V.A. Zalgaller, on the cutting of materials by linear programming methods. However, the transition from 2D objects to 3D objects

a) *b)*

Figure 4.
Aircraft about the same takeoff weight of different eras: (a) 30 years of the last century and (b) current.

and the complexity of the shape of the placed objects from linear strips to real objects of modern technology caused an avalanche of complexity of the mathematical description of the placement process. In our opinion, the main efforts of scientists-geometers were directed to the study of various aspects of the computer description of the form of technical objects, which are alternative to many classical methods of descriptive geometry. Therefore, now there are no problems to describe the geometric shape of objects of almost any complexity with the necessary accuracy.

The development of methods for automated placement of objects according to specified criteria was much less fortunate. In placement problems, it is important not to accurately describe the geometric shape of the objects being placed, as to solve two critical issues:

- Detection of cases of mutual intersection of the placed objects.

- Generation of options for placing objects in a given space, providing an effective layout. The implementation of the known algorithms of automated placement is based on polygonal models and computational methods of "brute force," which does not allow them to be implemented in practice for objects of complex structures, even with the modern power of computer equipment.

It can be objected that the use of modern CAD systems allows to carry out modeling of quite complex objects and at the same time to track possible cases of their crossing by means of the CAD system (**Figure 5**). But in this case, we are not talking about computer-aided design, but only about checking the layout variant already generated taking into account the experience and intuition of the designer.

The problem of computer representation of geometric objects of any complexity can now be considered sufficiently solved. However, other requirements are imposed on automated layout models, of which the accuracy of the form description is not the main thing. We have to choose which is better—an accurate geometric model, automatic layout of which is impossible, or a coarser geometric model, but allows the possibility of automated layout.

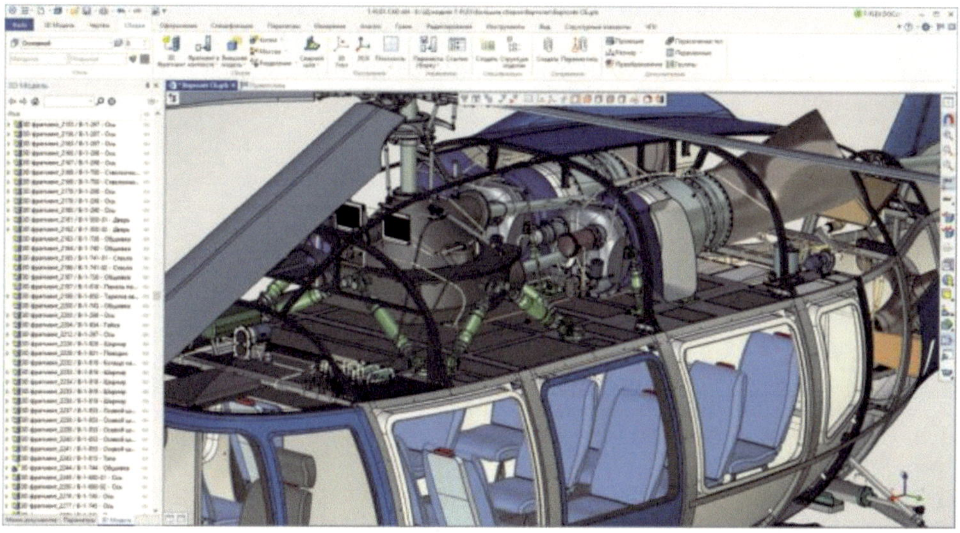

Figure 5.
Modeling layout in CAD systems.

3. Discrete (voxel) geometric models

It is known that the most accurate formal description of a three-dimensional object as a geometric body is its identification with the area of space occupied by it (a point set). However, in this formulation, the problem of the formation of a geometric object (GO) can be considered only theoretically. This concept can be used in practice if as the initial element of the set (E^3) take not an infinitesimal point but, for example, a cube with dimensions ($l \times l \times l$). In this case, the condition of congruence of all cubes filling the space must be satisfied, and any two cubes must not have common internal points.

The space in this case is called discrete, and the geometric model formed in such a space, respectively, is discrete or voxel model. The term voxel formed from the words volumetric and pixel—elements of the three-dimensional image. Voxels are analogous to pixels in a two-dimensional to three-dimensional space. Voxel models are often used for visualization and analysis of medical and scientific information, as well as in computer graphics (most often in games (**Figure 6**)). Polygonal models inside are empty (and often this is enough—why do we, for example, know what is inside a computer character?). Voxel models are completely filling their insides—volumetric voxel cubes, which can contain additional information about the object.

The discrete method of geometric modeling (in relation to technical applications) was described in the early 70 years of the last century by the Belarusian scientist, Zozulevich in [4], but in those years it did not spread due to the limited capabilities of computers in memory and performance. Although he and the team of his employees solved certain applied problems using this method, it was impossible to count on the effective use of voxel models with computers of those years with 16-bit architecture and 32–128 kb of RAM.

The basis of such models is an approximate representation of a geometric object in a field or voxel space. For the flat case, the voxel field is a uniform rectangular network $m \times n$, each cell of which is considered as a separate voxel, which can have two states—"0" or "1." Mathematically, the voxel geometric model is described by

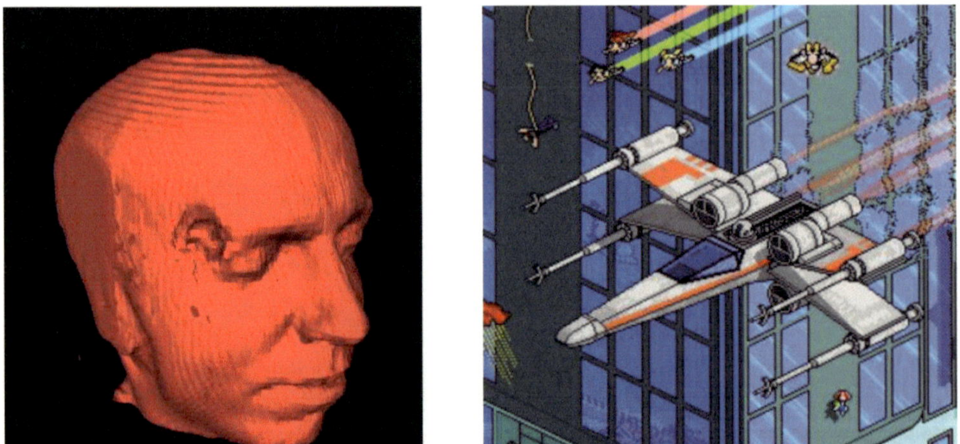

a) *b)*

Figure 6.
Voxel models in medicine ((a) the result of computed tomography) and in computer graphics ((b) in computer games).

the set $A = \{a_{i,j}\}$, where the voxel is considered unexcited if the object boundary does not pass through it and it does not belong to the inner region (**Figure 7**).

3D objects are described by a three-dimensional matrix $A = \{a_{i,j,k}\}$ with dimension a of $m \times n \times p$ (**Figure 8**). Obviously, the accuracy of the description of the geometric shape of the object depends on the chosen discreteness of the voxel matrix.

The author of the method Prof. Zozulevich D.M. called his method "receptor" by analogy with the receptors of the human brain; each of which can be either excited or not. Other names of this method are also known—"matrix," "binary," "enumeration of space elements," etc. In its geometric essence, the receptor (voxel) method is a special case of the analytical approximation method, which is used to describe three-dimensional objects that include complex surfaces of the second and higher orders. Since computational processing of such surfaces is difficult, they are approximated by areas of surfaces of lower order (planes, cylinders, etc.).

Research and development of voxel geometric models for various applications was carried out in the works of Russian scientists Gorelik AG, Gerasimenko YP, Klishin VV, Romance YA, Pashchenko OB, and Toloka AV, as well as a number of foreign authors—Gargantini I, Requcha AAG, Si Thu Lin, Nyi Nyi Htun, Kyi Min Han, Ye Win Tun, and a several others.

Here, it should be noted that very close in research ideology of Nazarova KM, Ratkova SI, etc., in which as basic object shape is not classic receptor in the form of a cube or parallelepiped, and more complex shapes—for example hexahedron.

The voxel method has both advantages and disadvantages. The main advantage is the homogeneity of computational algorithms and a very simple detection of cases of intersection of objects with each other.

The obvious disadvantages include the discreteness of the model and the need for large amounts of computer memory for its implementation. However, now the increase of computer memory to any volume is neither technical nor economic.

Another drawback is that the voxel geometric model is never primordial. Placed and already placed products are described by the designer, usually parametric geometric models (i.e., specifying the type of object and its parameters—a sphere with radius R, a parallelepiped with dimensions $a \times b \times c$, etc.). Examples of parametric models are shown in **Figure 9**. Therefore, there is a need for an additional software module "parametric model"↔"voxel model." However, now there are ways to form a voxel matrix directly on the solid-state model created in any of the CAD systems (**Figure 10**).

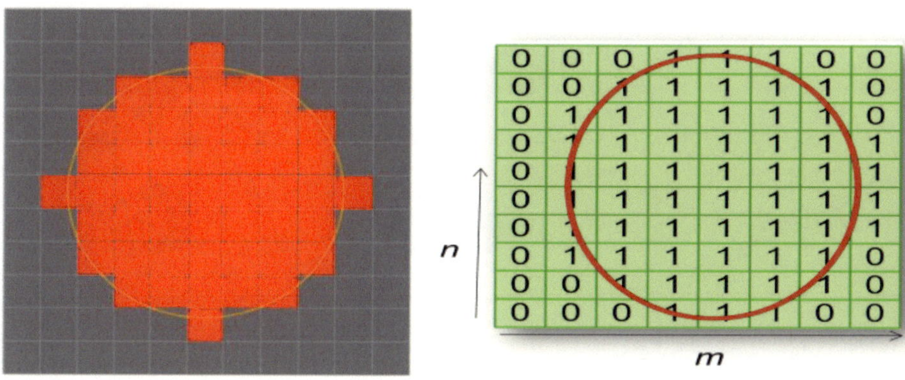

Figure 7.
Voxel model of a 2D object.

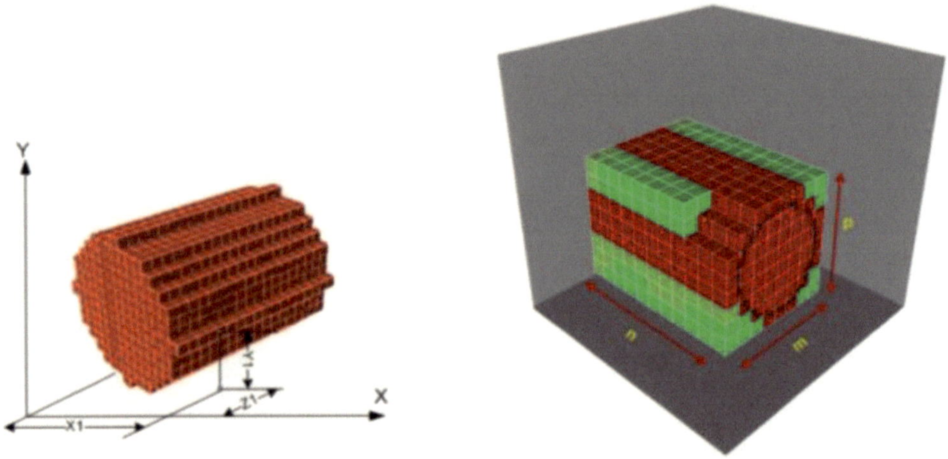

Figure 8.
Voxel model of a 3D object.

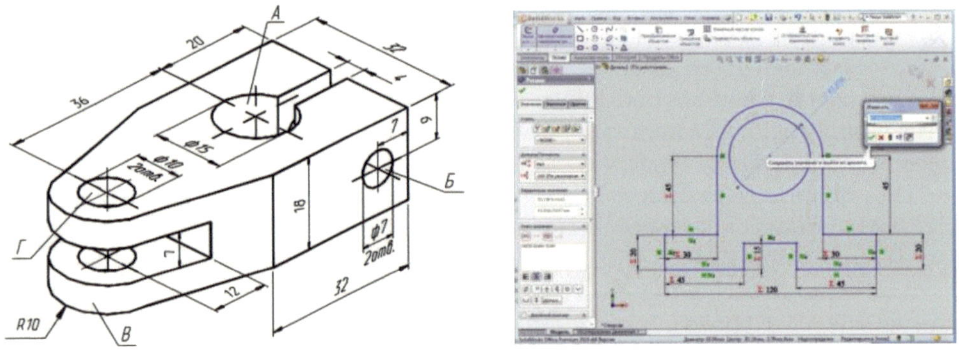

a) *b)*

Figure 9.
Setting parameters of geometric objects in traditional drawings (a) and computer modeling (b).

a) *b)* *c)*

Figure 10.
Stages of construction of voxel model of an electric drill: (a) initial drawing, (b) solid model from CAD system, and (c) voxel model.

4. Examples of computer-aided design problems using voxel geometric models

4.1 Problem "rearrangement"

Consider the problem of placing additional equipment in the technical compartment of the vehicle (the problem of pre-assembly). This problem is often found in the practice of design and from a geometric point of view is reduced to the problem of analyzing the geometric shape of empty spaces among previously placed objects.

We will solve the problem of additional placement in the following statement—there is a closed area (e.g., the technical compartment of the aircraft) with the equipment already partially placed in it (**Figure 11a**). There is a set of equipment that needs to be "repositioned" (**Figure 11b**). The possibilities of "additional placement" are determined by the shape and size of still unfilled spaces between already composed objects.

For a person, such a task "re-arrangement" of objects will seem quite easy—he can easily "by eye" determine how several volumes relate to each other and which object fits in another and which does not. To do this, he mentally classifies the object by shape (almost a ball, almost a cylinder) and then mentally correlates their sizes. Unfortunately, this operation of pattern recognition, which is so simple for a person, is of considerable complexity even for modern computers.

The most important issue is the class of geometric objects allowed to be placed. Typically, from a geometric point of view, such equipment is either primitives or a combination of primitives. Assume that the objects to be placed are a composition of primitives that describe the shape of the instrumentation quite well (this can be seen in **Figure 11b**). The technical literature provides data that the composition of primitives effectively describes 95% of the instrument equipment.

The advantage of the voxel geometric model is the ability to identify empty spaces in the layout for subsequent placement of objects not yet placed in them. If there is an empty space among the assembled objects, then on the corresponding section of the voxel matrix it is relatively easy to identify as "clumps" of zero voxels. For this purpose, an algorithm was developed to determine the center of a certain free area (flat) and its dimensions.

To do this, scan the rows and columns to identify the largest cluster of inextricably located receptors and identify their centers. To analyze the individual cross

a) *b)*

Figure 11.
Technical compartment of the aircraft: (a) "redeployable" and equipment (b).

section of the object, we use the transition from the geometric shape of the object (in the form of a discrete set of receptors) to the "feature space" adopted in the theory of pattern recognition. Such a feature space for us will be the hodograph of the function of the radius of the vector from the center of the section (**Figure 12**):

$$F = R_i(\varphi_i)$$

where R_i is the current radius-vector length for the i receptor and φ_i is the current radius-vector angle for the ith receptor.

After constructing the function R_i (φ_i), its analysis begins. If the shape of this region were a circle, the function would be an ideal straight line whose height would show us the radius of this circle (**Figure 12b**), if a regular polygon is a "saw" with the number of vertices equal to the number of sides. The coordinates of the vertices on φ will determine the aspect ratio of the rectangle. To do this, we use the method of testing statistical hypotheses, implemented as a computational software module.

On the real results of the analysis of the hodograph function, naturally superimposed "noise" due to the discreteness of the description of the component objects. An example of such real hodographs of a plane slice of the objects being assembled, statistically identified as slices of "polygon," "cylinder," and "sphere," are shown in **Figure 13a–c**, respectively.

An illustration of the software implementation of this method is presented in **Figure 14**. The obtained results are implemented in the framework of a graphical shell written in C#. If we have a desire to place in this area not a parallelepiped of the maximum size, but, for example, a sphere of a certain radius, then it becomes a full participant of the scene, and after pressing the "analysis" button, a new definition of the configuration of unfilled spaces occurs.

The solution of the task was carried out within the framework of the dissertation research by Situ Lin (Republic of the Union of Myanmar), a postgraduate student of MAI and described in [5].

4.2 Task of tracing channel

The purpose of designing the channel surface is the delivery of a certain trajectory of a material carrier (liquid, gas, electrical energy from one point (entry point) of the technical product to another (exit point)). In this case, one of the tasks of the layout is to solve the problems of tracing, i.e., designing communications between

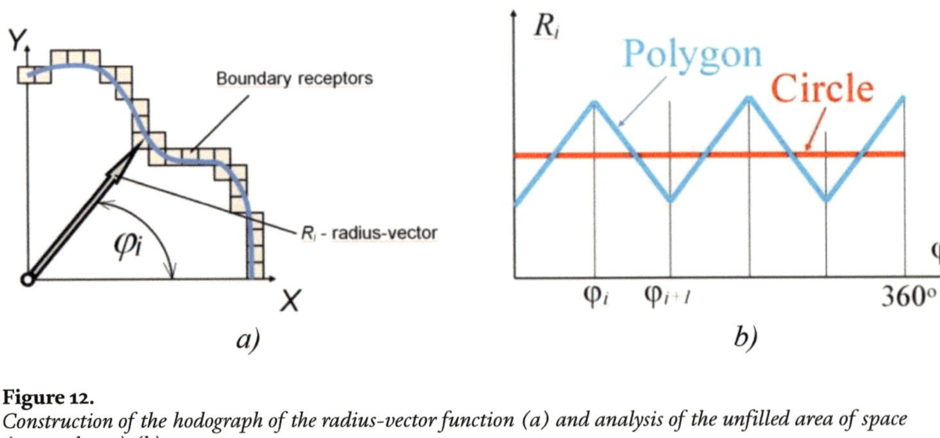

Figure 12.
Construction of the hodograph of the radius-vector function (a) and analysis of the unfilled area of space (case-polygon) (b).

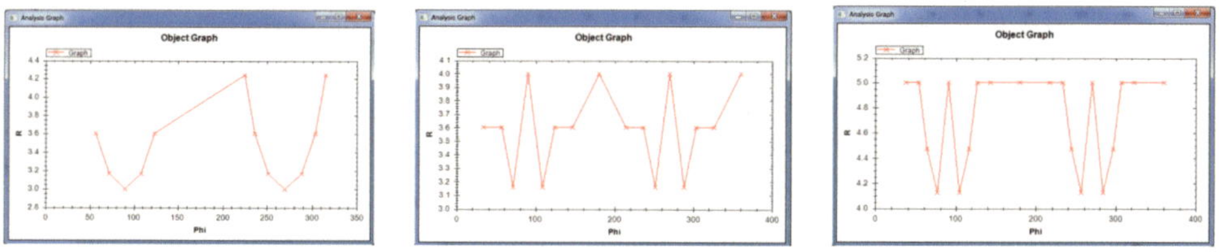

a) *b)* *c)*

Figure 13.
View of the hodograph of the slice function from for the polygon (a), sphere (b), and cylinder (c).

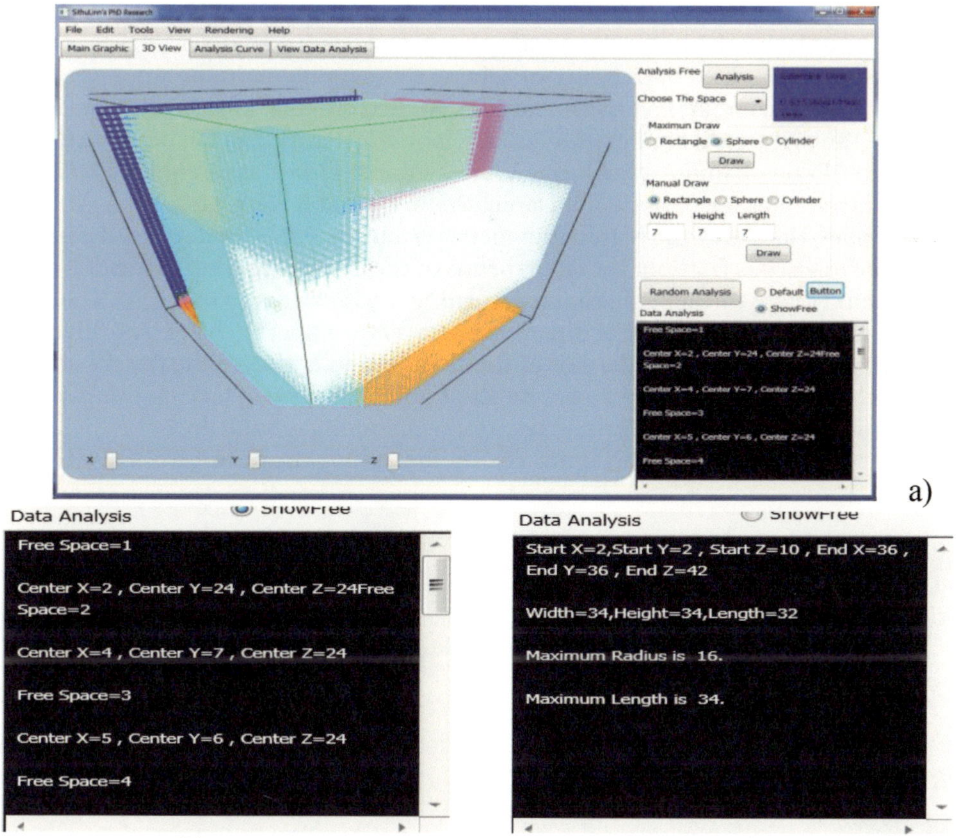

a) *b)* *c)*

Figure 14.
Determination of the configuration of empty spaces: (a) visualization of space, (b) output of data on the size and position of the free area, and (c) assessment of the possibility of inscribing different shapes.

already placed objects. Such problems are quite difficult to formalize and difficult to solve because of their inherent multi-extreme nature.

A special and much more complex type of trace is called "solid" trace, that is, such a case when the dimensions (connecting elements) of the trace are comparable to the dimensions of the components. In practice, this is the design of pipelines, air ducts, and other elements of transport systems (**Figure 15**).

To solve this problem, two main approaches to tracing are used, which are determined by the metric used. A metric is a rule that determines the distance between two points in a given space. The first approach is to use the Euclidean metric. In this case, the trace is drawn in the direction of the shortest distance between the entry and exit points (**Figure 16a**), and the length of the trace is determined by the Pythagorean theorem.

The second approach is to use the Manhattan metric, in which the trace is drawn in the direction of the coordinate axes (**Figure 16b**). In this case the trace is much longer than using the Euclidean metric, but the approach itself provides additional mathematical possibilities for the design of the trace. It is used for tracing large integrated circuits and printed circuit boards; this metric is used by industrial automated tracing systems P-CAD, TopoR, and others. A typical example of the result of wiring in such programs is shown in **Figure 16c**. From this figure it can be seen that the tracing in such systems is carried out either by the Manhattan metric or at 45° angles.

In our opinion, the problems of trace design according to the Manhattan metric are the most developed in the theory of geometric modeling of traces. This is due to the extreme practical importance of solving these problems in the automated wiring of printed circuit boards and large integrated circuits. Fundamental in this field are the theoretical studies of Russian scientists Abraitis LB, Bazilevich RP, Petrenko AP, Tetelbaum AY, Selyutina VA and others, as well as foreign scientists EW Dijkstra, Judea Pearl, Ira Pohl, Daniel Delling, Peter E Hart, etc.

However, the use of the Manhattan metric is unacceptable for channel design, since due to viscous friction, the sharp bends of the channel make it difficult to move liquid or gas through the channel, turning the flow energy into heat. The current line that determines the direction of travel through the channel is called the main guide line (GNL). It is the axis of the channel and is given either by the equation of the spatial curve or by a discrete set of points. However, tasking this parameter alone is not enough—it is necessary to specify the shape and area of individual cross sections of the channel (**Figure 17**). By controlling the position of the GNL

a) *b)*

Figure 15.
An example of laying the trace: (a) electric furnace and (b) the air duct channel of the car.

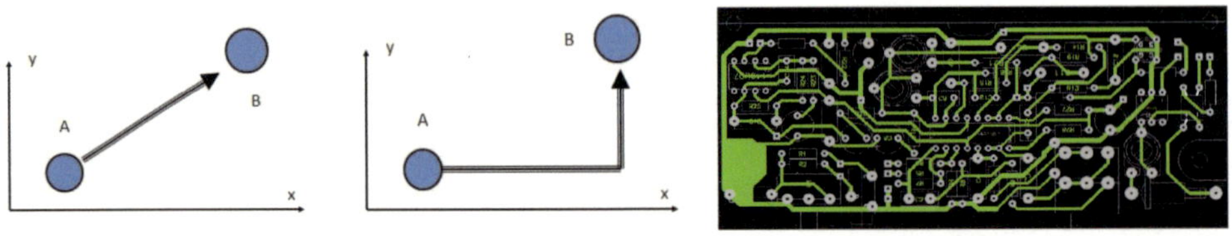

(a) (b) (c)

Figure 16.
Designing trace with different metrics.

and the shapes and sizes of the cross sections, we can provide the specified designer characteristics of the flow of liquid or gas through the channel.

Let us complicate the task of designing. If earlier the channel was first designed according to the specified characteristics and then it was already placed, then we are trying to design a channel with the specified characteristics, "inscribed" in an already existing layout. Let there be a rectangular area with dimensions $X \times Y$, in which the areas of prohibition are located. The entry point A and exit point B of the channel are specified (**Figure 18**). It is necessary to draw a trace between the given start and end points A and B. Obviously it is that in such a statement the problem does not always have valid solutions.

From **Figure 18** it can be seen that there are quite a lot of options for its passage. At first glance, of all the traces in **Figure 18**, the one that is shorter will be better. However, it is known from design practice that hydraulic losses are minimal not in the short but in the main channel. Additional requirements are possible, for example, to provide a specified gap between the channel and other elements of the layout.

A significant problem in solving the tracing problem is to avoid obstacles, which are already composed objects or communications between them. The big advantage of the voxel approach is the ease of detecting an obstacle by voxel code (0 or 1). The simplest approach is to ignore obstacles before encountering them. Such an algorithm would look something like this: choose the direction to move toward the target, and move until the target is reached and the direction is free to move (**Figure 19**). Given that the transverse area of the channel can repeatedly exceed the size of one voxel, it is possible that a valid trace option in this layout situation is not at all.

The known trace algorithms closest to our approach are described in the following [6, 7]:

- Dijkstra's algorithm
- Algorithm A* "A the asterisk"

The simplest approach implemented in these algorithms is to ignore obstacles before encountering them. The structure of the choice of direction of movement is determined by the rule to move back to choose a different direction in accordance with the strategy of the traversal.

This allows us to argue that these algorithms contain elements of artificial intelligence (AI), since the solution is chosen according to the predicative principle of "If"-"Then."

In the works of Dijkstra EW, Donald Ervin Knuth, Thomas H Cormen, etc., various obstacle avoidance strategies (heuristics) based on both random search and artificial intelligence algorithms are analyzed. Each of them has both its limitations and areas of preferred application. Examples of different trace algorithms are shown in **Figure 20a**. **Figure 20a** shows that although the well-known Dijkstra algorithm was able to pave the way from the start to the end points, it did not do it rationally, making an extremely many unnecessary movements and repeatedly unnecessarily changing the direction of travel.

Voxel algorithm A* operates in a more reasonable way (**Figure 20b**) and is commonly used to find the optimal shortest path. But this algorithm is not able to take into account the given gap (δ-neighborhood)—the route may pass too close to the areas of prohibition. Thus, the search for the path of the trace by the algorithm A* gives the best results but does not always provide a solution to the problem. In

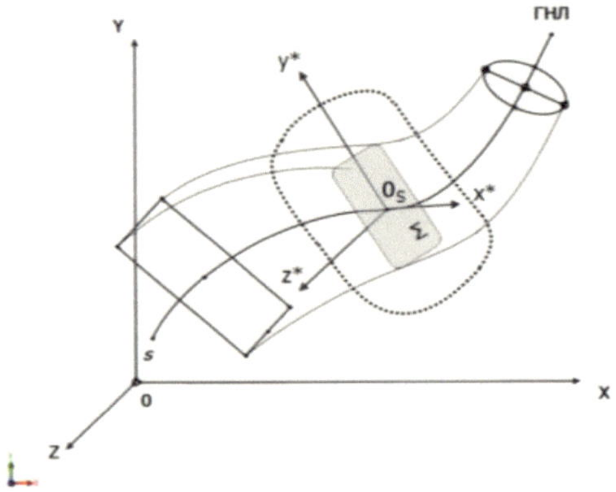

Figure 17.
Geometric parameters that determine the shape of the channel.

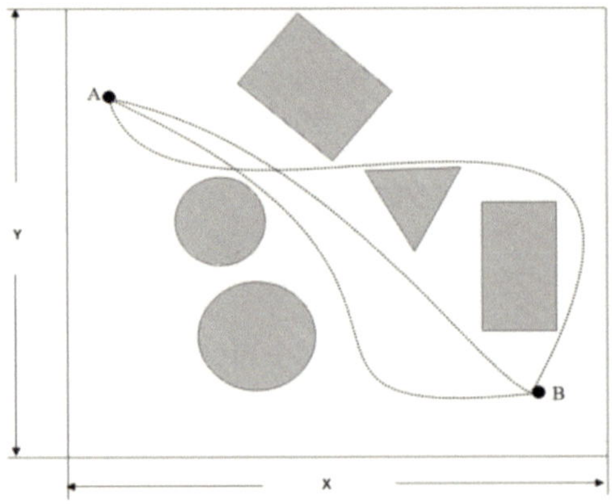

Figure 18.
Finding a rational path between two endpoints A and B in a 2D formulation without taking into account the size of the trace and the restrictions on smoothness.

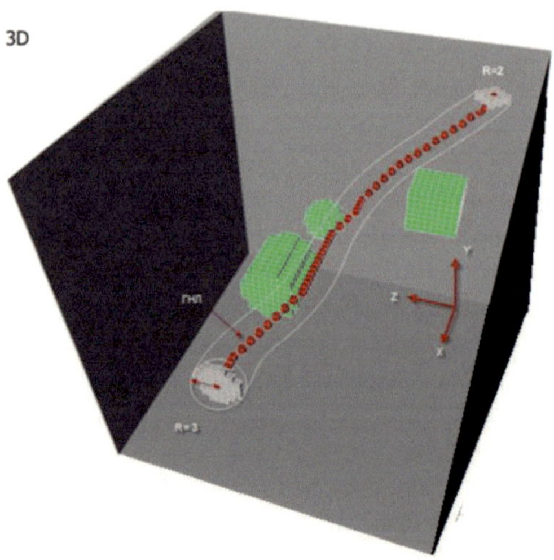

Figure 19.
The principle of obstacle avoidance in the construction of voxel method.

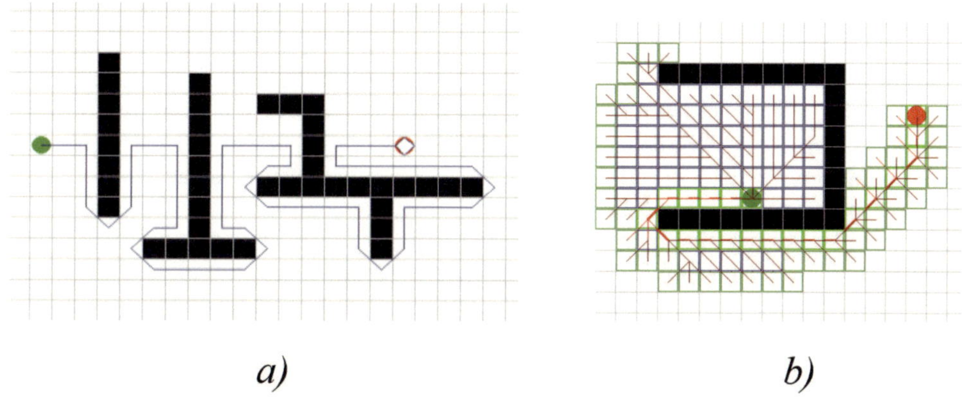

Figure 20.
Examples of tracing when bypassing various obstacles in 2D staging.

addition, the algorithm A* is not able to design trajectories with a given degree of smoothness, since such a task has never been set before.

As it was already noted earlier, the strongest side of the algorithm A*, which led to its choice by us as a prototype, is the possibility of optimal and heuristic trajectory search. The literature shows that in many cases heuristic search works better than other search strategies. The heuristic function of the algorithm determines the choice of the search direction to the target vertex. If the heuristic function is valid (that is, does not exceed the minimum cost of the graph to the target vertex), then the algorithm A* is guaranteed to find the shortest path. The modified algorithm A* uses a set of heuristics that provide for multidirectional search. The search for direction is conducted not as usually on 4 and 8 directions (respectively in a plane and space (**Figure 21a** and **b**)) and in 8 directions, if the construction of the channel occurs in the plane and 26 adjacent vertices in the design of the spatial channel (**Figure 21c** and **d**).

To implement the proposed trace algorithm, the Advanced Pathfinder System (APS) program was written in Microsoft Visual Studio 2010, using the C# programming language. The solution of the task was carried out within the framework of the dissertation research by Nyi Nyi Htun (Republic of the Union of Myanmar), a postgraduate student of MAI and described in [8]. With this program, you can:

1. Use the improved algorithm A* and find a rational trace between two points in 2D and 3D spaces, taking into account the areas of prohibitions.

2. Carry out smoothing of the trace received at the previous stage either on any set radius or on the maximum possible radius with the subsequent check of accomplishment of a condition that the minimum radius is not less than the set R_{min}.

3. Ensure that the trace passes at the specified minimum distance δ from already placed objects and areas of prohibition.

Testing showed significantly higher performance of our modified algorithms than the original one. It should be recognized that the described program, which implements the voxel method of body tracing, is not integrated into any existing CAD system, so the geometric information is entered into the program in parametric form. This reduces the performance of the build process. Integration of this program with any CAD system in the form of a separate built-in calculation module is an actual technical task.

4.3 Task of solar layout

When designing the SC, the question on estimation of the effective area of solar panels arises, taking into account their inevitable shading by each other and other structural elements of the spacecraft SC (**Figure 22**). All this significantly limits the functionality of the SC. When designing SC or ground-mounted solar power plants, we have to decide on the area of solar panels—if there are a small number of panels, then the solar energy absorption will be small, and if there are too many, they will work inefficiently, shading each other (not to mention the additional costs for them and increasing the mass of the entire SC). Therefore, the solution of this question can be considered as an optimization problem of mathematical programming.

Voxel geometric models do not require complex formulas or logical constructions for their implementation. However, their practical implementation has its own specific complexity. In addition to the need to convert the initial parametric model specified by the constructor into a receptor model, the complexity is conditioned by the need to take into account the position and value of each voxel (out of many millions in the voxel matrix), as well as to create a mechanism for visualizing the results. The solution of the task was carried out within the framework of the dissertation research by Kui Min Khan (Republic of the Union of Myanmar), a postgraduate student of MAI and described in [9].

An essential feature of our approach is that we will not use the classical voxel matrix (filled with "0" and "1") in the calculations, but a ***multi-digit*** one, in which additional codes will be added. Specifically, it will be three-digit—"0," free space, "1," space occupied by the space station residential module, and "2," space occupied of solar panels (**Figure 23**).

Using a multi-digit voxel geometric model allows you to proceed directly to the calculations of shading. We will move a slice of the voxel matrix with thickness of 1

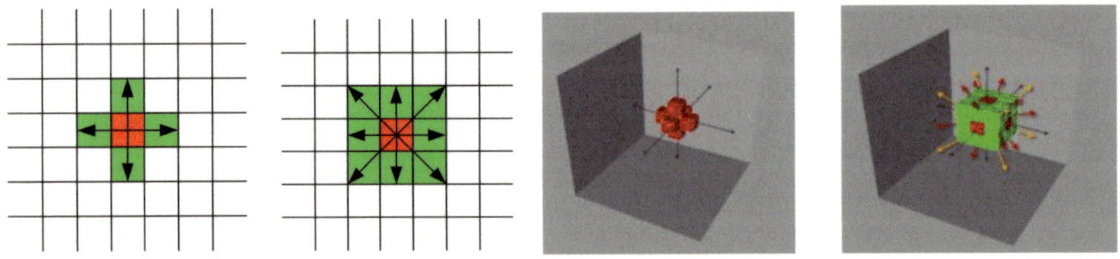

a) *b)* *c)* *d)*

Figure 21.
Increasing the directions of the path search in the original and modified algorithms.

Figure 22.
Partial shading of solar panels in space on the International Space Station (ISS).

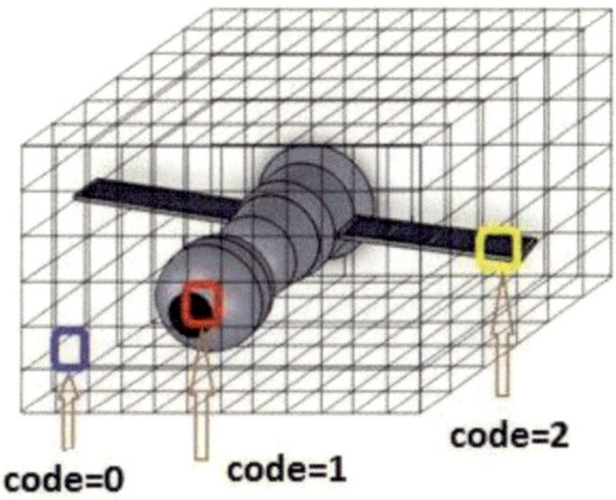

Figure 23.
Representation of SC by multi-digit voxel matrix.

Figure 24.
Single-layer slice of the voxel matrix (a), moving this slice along the voxel matrix (b).

voxel (**Figure 24a**) as a cutting plane along the coordinate plane from the beginning to the end of the voxel matrix (**Figure 24b**).

In **Figure 25a**, it can be seen that on each slice we can put each specific voxel in accordance with either "1" (if it coincides with the body of the SC) or "2" (if it coincides with solar panels). If there is no match with any elements of the SC, the value of the voxel on the slice remains the initially set value of "0." Looking at a single-layer slice of the voxel matrix from the direction of the energy flow *W* (**Figure 25b**), we see the layer filled with nonzero code that allows to calculate the total area of space (number "0"), the total area of the habitable modules of the SC (number "1"), and, most interestingly for us, the total area of solar panels (number "2"). Next, everything seems to be simple—summing up the area of voxel with codes "2" on all sections, we get the area of unshaded zones of solar panels.

In this calculation model, there are situations when along the thickness of the solar panel may be not one, but several layers of the voxel matrix (for example, 4 layers), resulting in the unreasonable increase of the effective area of solar panels by four times. It is also necessary to exclude unreasonable repeat account of already screened objects. To do this, we introduce an additional code "3" in the voxel matrix, which will exclude the account of the areas of the corresponding voxels. The essence of the model change is that once absorbed part of the energy flow should no longer be taken into account. Therefore, starting with some slice of the voxel matrix, everything that follows this slice, the element with the code "2," is forcibly

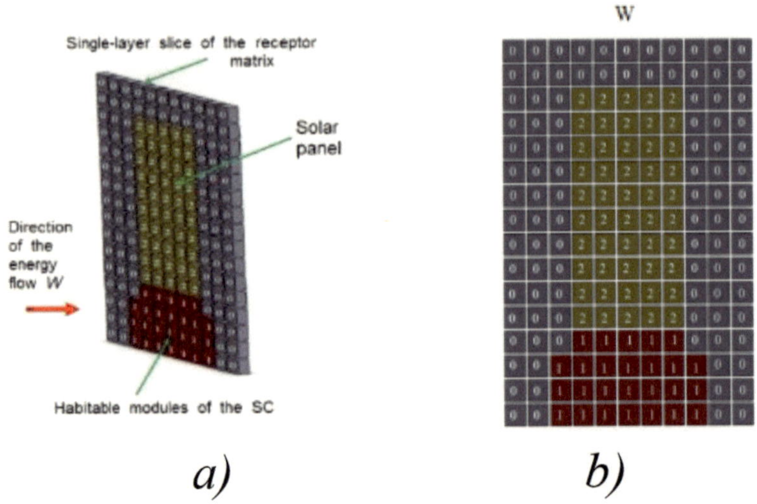

Figure 25.
Single-layer slice of the receptor matrix (a), the view of this slice toward the flow (b).

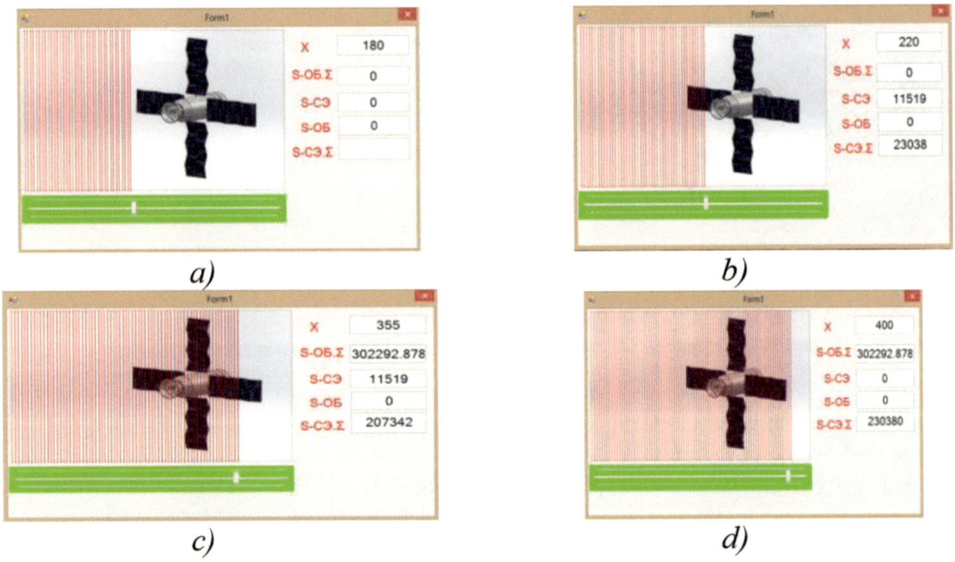

Figure 26.
Scanning stages of 3D model SC with inclined position for calculation of sectional areas of solar panels and SC body.

filled with the prohibiting code "3," which does not allow the use of voxels with this code in any calculations.

However, the shading of solar panels in the *W* direction, reducing their efficiency, is possible not only by other solar panels but, in some cases, by other elements of the SC (e.g., the body). Therefore, we make another change in our model—filling the entire voxel matrix in the direction of *W* with codes "3" after the first detection on the slice of any element of the SC. As in the previous case, the entire remaining part of the voxel matrix in the direction of the energy flow *W* is filled with codes "3," which excludes the participation of voxels with this code in the calculation of the effective area of solar panels *S*. Therefore, in the modified (4-digit) voxel model, voxels with the code "3" do not participate in any area calculations.

On the basis of the geometric model described above, a software package was created, implemented in the C# language, allowing to simulate the effective area of solar concentrators. At the same time, a graphical shell is developed that visualizes the calculation process and the calculated parameters of the effective area.

The work of the software package is as follows. After entering the information about the geometric dimensions of the station and solar panels (in parametric form), a layer-by-layer scan of the sections begins. A 2D matrix is formed in each layer, the form of which was previously shown in **Figure 24b**, from the 3D matrix, in which our entire object (SC) is immersed. In each section (slice) of the voxel matrix, the areas of the current section of the solar panels, the effective (cumulative) sectional area of the solar panels, and the cumulative sectional area of the body of the space station are calculated (although this parameter has no practical value for us). **Figure 26a** shows that the cutting plane of voxel matrix has not yet reached the model of SC, so all the sectional areas are zero.

Figure 26b shows that the cutting plane already passes on the SC itself, crossing both the solar panels and the SC body, so specific calculated values are obtained in each section of slice area, which will be visualized in the corresponding program windows. And finally, Figure *c* shows that the cutting plane completely passed through all 3D model of SC, therefore both current and the cumulative sums of the areas will not change any more in the program windows. Thus, we have solved the task—to determine the total visible area (from a certain angle) both of the body of the space station and its solar panels.

5. To evaluate the accuracy and efficiency of voxel model

The advantage of the voxel method is the simplicity of determining many geometric parameters and solving many geometric problems of processing three-dimensional objects. However, when using voxel model, the shape of the object is described approximately. The desire to increase the accuracy of the description in r times leads to an increase in the necessary computer memory for the implementation of the method in times where q is the dimension of the modeling space.

To determine the appropriate accuracy of the representation of 3D objects voxel matrix as a test example, consider the most unfavorable for the accuracy of the description of the case of the ball. Let us assume that a ball with a radius of 500 mm is inscribed in a voxel matrix with a variable number of voxel, which varies in the range from 1000 to 25,000 (**Figure 27a**).

For **Figure 27b**, the results of the calculation of the value of the most probable absolute error depending on the number of voxels for this matrix with dimensions 1 m × 1 m × 1 m are given. It can be seen from this graph that the error of the voxel model shape description is expected to depend on the voxel size, which, in turn, is determined by the number of voxels in the matrix. There is a threshold (in our case it is about 5000 voxels), with a decrease in which the absolute accuracy of the description of the form is sharply reduced. In [10] it is shown that for the most typical sizes of technical objects, even with a not very high number of voxels (about 5000 on each axis), the relative accuracy of the description remains quite high (below 0.05%), which is quite enough for most technical applications.

Since receptor models are discrete by definition, the discussion of the results raises the question not only about the accuracy of their description but also about the computational resources required for their implementation. Studies show that with a seemingly huge number of computational operations, they are performed surprisingly quickly. This is probably due to the homogeneity of the calculations and the use of only the RAM of the computer for their execution. The time of

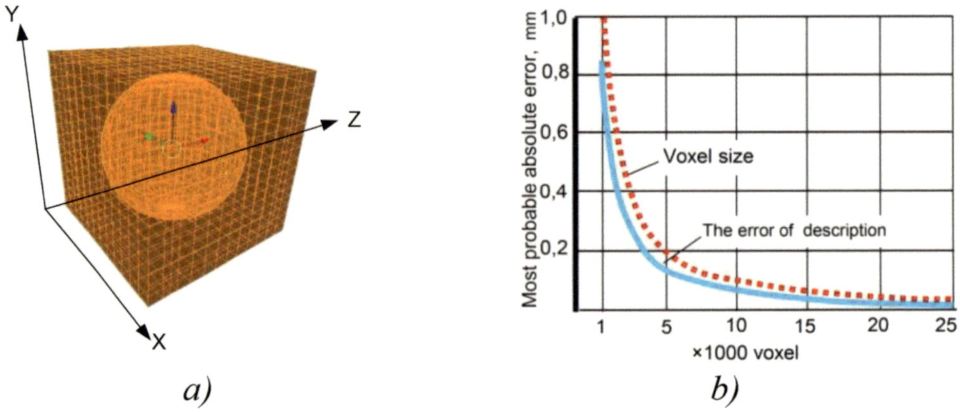

Figure 27.
Results of calculation of the error of the shape description by the voxel matrix.

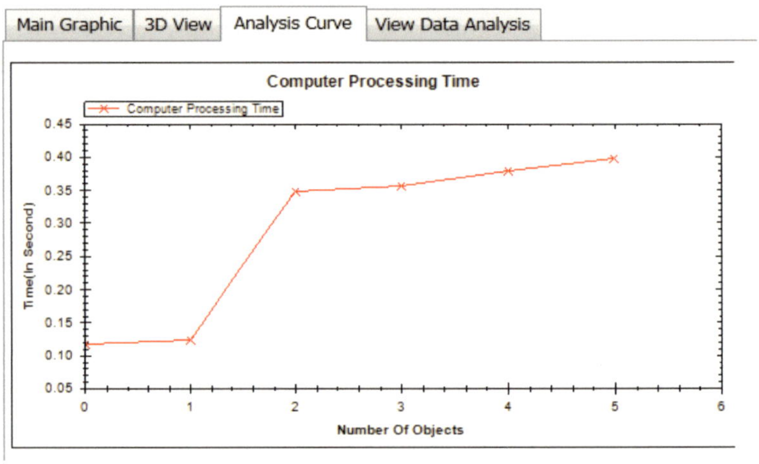

Figure 28.
CPU time required to determine empty spaces depending on the number of objects already placed.

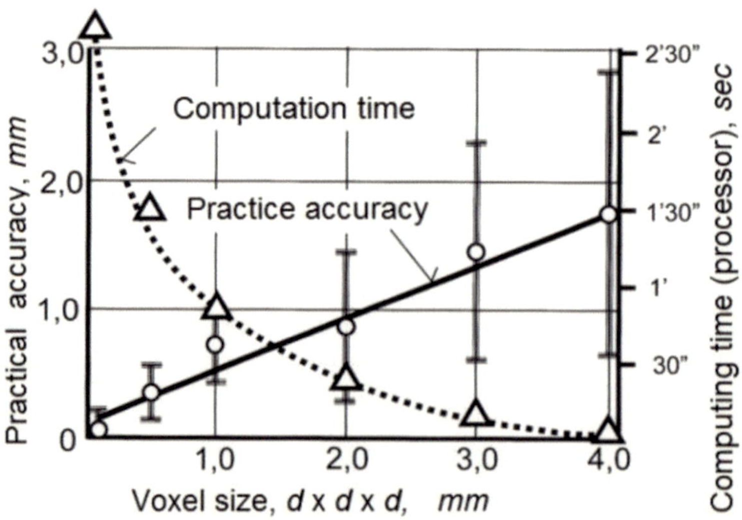

Figure 29.
Calculated characteristics of voxel geometric model accuracy in designing a channel between obstacles.

computer processing of voxel models is determined by both the hardware capabilities of the computer and the parameters of a particular scene—the number of voxels and the number of objects already placed. In **Figure 28** the estimation of processor time in solving the problem of recognition of a flat area with a fixed number of voxels (1000 × 1000) is given. **Figure 28** shows that the CPU time is still relatively small (fractions of a second), even for a medium-power computer. For a spatial scene, the time to obtain a solution increases by about an order of magnitude, but still remains acceptable for interactive mode. But it should be borne in mind that for voxel geometric models are possible methods of decomposition and parallelization of calculations, which will significantly speed up the calculation time.

Evaluation of the accuracy and performance of the channel design between obstacles in the test example using a voxel matrix size 1 m × 1 m × 1 m is shown in **Figure 29**. They also show acceptable practice accuracy and simulation time.

6. Conclusions

Discrete methods of geometric modeling, which include voxel, can effectively track the cases of intersection of simulated objects in space, which makes it advisable to use them in automated layout problems. Voxel models allow you to create intelligent algorithms for automated placement. This eliminates the need to use multiple methods of search of accommodation options.

The unsolved problem of using voxel geometric models in the problems of automated layout is their integration into modern CAD systems, which limits their widespread introduction into the practice of computer-aided design.

Author details

Leonid V. Markin
Moscow Aviation Institute (National Research University), Moscow, Russian Federation

*Address all correspondence to: markinl@list.ru

References

[1] Osipov VA. Mashinnye metody proektirovaniya nepreryvno karkasnyh poverhnostej [Machine Methods for Designing Continuous-Frame Surfaces]. Moscow: Mashinostroenie Publishers, 1979. 248 p

[2] Markin LV. O putyah sozdaniya geometricheskih modelej avtomatizirovannoj komponovki [On the Ways of Creating Geometric Models of Automated Layout]. Geometriya i grafika [Geometry and Graphics]. 2015;**3**(1):64-69

[3] Deniskin YI, Egorov EV, Nartova LG, Kuprikov MY. Prikladnaya Geometriya. Nauchnye Osnovaniya i Primenenie v Tekhnike [Applied Geometry. Scientific Grounds and Application in Technology]. Moscow: MAI Press Publishers; 2010. 385 p

[4] Zozulevich DM. Mashinnaya Grafika v Avtomatizirovannom Proektirovanii [Computer Graphics in Computer Aided Design]. Moscow: Mashinostroenie Publishers; 1976. 240 p

[5] Lin S. Razrabotka Metodov i Geometricheskih Modelej Analiza Nezapolnennyh Prostranstv v Zadachah Razmeshcheniya [Development of methods and geometric models for the analysis of unfilled spaces in placement problems] [thesis]. Moscow Aviation Institute: Moscow; 2011

[6] Dijkstra EW. A note on two problems in connexion with graphs. Numerische Mathematik. 1959;**1**:269-271

[7] Khantanapoka K, Chinnasarn K. Pathfinding of 2D & 3D game real-time strategy with depth direction A* algorithm for multi-layer. In: Eighth International Symposium on Natural Language Processing, Thailand. 2009. pp. 184-188

[8] Htun NN. Razrabotka i issledovanie receptornyh geometricheskih modelej telesnoj trassirovki [Development and research of receptor geometric models of body tracing] [thesis]. Moscow: Moscow Aviation Institute; 2014

[9] Khan KM. Matematicheskoe i programmnoe obespechenie rascheta zatenennosti solnechnyh batarej kosmicheskih letatel'nyh apparatov [Mathematical and software for calculating the shading of solar panels spacecraft] [thesis]. Moscow: Moscow Aviation Institute; 2018

[10] Khan KM, Markin LV, Tun EV, Korn GV. Receptornye Modeli v Zadachah Avtomatizirovannoj Komponovki Tekhniki [Receptor Models in the Problems of Automated Layout Technology]. Saarbryuken: Lambert Publishers; 2016. 110 p

Chapter 3

Effective Pitch Value Detection in Noisy Intelligent Environments for Efficient Natural Language Processing

Damjan Vlaj, Andrej Žgank and Marko Kos

Abstract

The performance of applications based on natural language processing depends primarily on the environment in which these applications are applied. Intelligent environments will be one of the major applications used to process natural language. The methods for speaker's gender classification can adapt and improve the performance of natural language processing applications. That is why, this chapter will present an effective speaker's pitch value detection in noisy environments, which then allows more robust speaker's gender classification. The chapter presents the algorithm for the speaker's pitch value detection and performs the comparison in various noisy environments. The experiments are carried out on the part of the publically available Aurora 2 speech database. The results showed that the automatically determined pitch values deviate, on average, only by 8.39 Hz from the reference pitch value. A well-defined pitch value allows a functional speaker's gender classification. In this chapter, presented speaker's gender classification works well, even at low signal to noise ratios. The experiments show that the speaker's gender classification performance at SNR 0 dB is higher than 91% when the automatically determined pitch value is used. Speaker's gender classification can then be used further in the processes of natural language processing.

Keywords: intelligent environment, pitch, speech processing, gender classification

1. Introduction

Human-computer interfaces (HCIs) are frequently those parts of modern information and communications technology (ICT) systems, which play a crucial role in the case when the products are entering the market [1]. HCI is, from a user's perspective, perceived as the entity which can control the system's functionality and, thus, improve the quality of experience [2]. One of the ICT systems where HCI has made significant development progress in the last decade is the smart home and smart city solution [3, 4]. Human communication with these systems can be carried out in the form of the spoken interaction, which is the most natural and frequent modality for users. Some of the commercially available spoken virtual agents are Alexa by Amazon, Siri by Apple, Google Now by Alphabet, and Cortana by Microsoft. These commercial virtual agents support major languages but are lacking support for under-resourced languages [5]. Another shortcoming is the lack of support for real spontaneous and emotionally driven conversation, which could improve the quality of the experience further [2].

Spoken virtual agents are using various natural language processing (NLP) techniques to communicate efficiently with humans (**Figure 1**). In this case, the NLP algorithms depend on the text input, which is produced via automatic speech recognition (ASR). Besides the direct text input, the ASR and its submodules can provide the NLP with additional meta-information, which can be used to improve the virtual agents' response to the user's communication. Some categories of such meta-information are emotions, stress level [6], effects of spontaneous speech, speakers' change [7]. As an example, change of speaker influences dialogue modelling, which can be seen as an essential part of language generation with NLP approaches. Advanced ASR systems can apply spontaneous speech modelling to reduce the ratio of errors produced by a subsequent NLP system, which has to process such an error-prone spontaneous input. Another example is the case when the NLP system is a part of an eHealth solution, where changes in stress level alter the NLP response of the virtual agent directly, either in the way of triggering some relaxing scenarios or forwarding this data in the form of NLP-generated information to caregivers [3].

An important characteristic of the human speech signal is pitch value, which can be used for both of the cases of ASR-NLP interaction mentioned above [8]. Pitch value can be used as one of the parameters, obtained during the feature extraction, which is the first step of speech recognition. Particularly, tonal languages (i.e. Mandarin) are such where pitch value information plays a crucial role in an ASR system. Moreover, the pitch value can improve speech recognition accuracy significantly in the case of spontaneous and accented speech, which is common in real-life human interaction with virtual agents. Pitch value can also be included as part of meta-information for NLP approaches, as it can be used to estimate emotions,

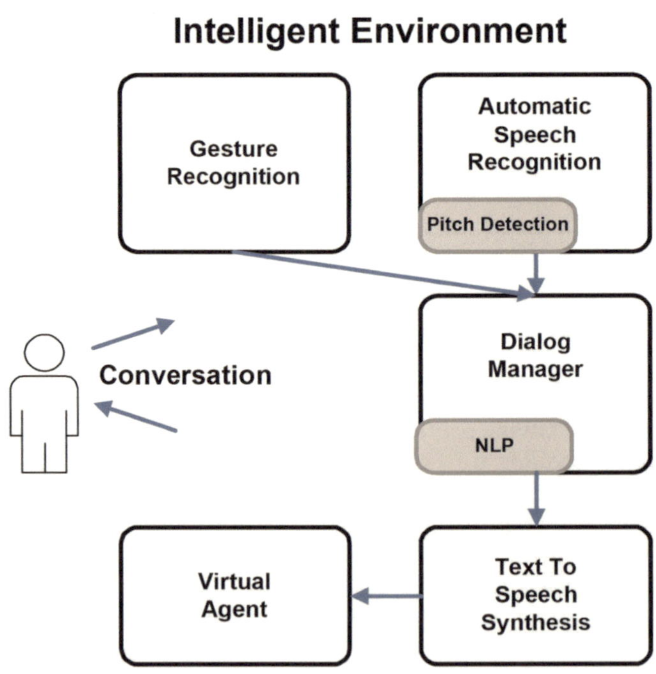

Figure 1.
Diagram of conversational human-computer interaction in an intelligent environment.

stress, and speaker information. The objective of this chapter is to present the importance of pitch value calculation from the point of view of automatic speech recognition (ASR) as a first building block of a natural language processing-based spoken virtual agent. Improved pitch value calculation can have a significant impact on the performance of an NLP system and, thus, also, in general, improve the quality of experience of virtual agents or other systems based on spoken human-computer interaction.

This chapter is organised as follows. The literature review is given in Section 2. This chapter will give a general view of the determination of the pitch value in the speech in Section 3. The pitch value determination method is presented in Section 4. The experimental design and results are given in Section 5. This chapter concludes with the discussion in Section 6 and conclusion in Section 7.

2. Literature review

Effective pitch value estimation (or fundamental frequency F_0, as it is also referred to) for various tasks and applications has been addressed for many years. It is one of the fundamental problems in speech processing because pitch value estimation is used in several different applications (e.g. speech recognition, speech perception, speech transformation, language acquisition, speech analysis, speaker identification) [9]. Good review work regarding pitch value extraction was presented by Gerhard [10]. Pitch value can be estimated in the time, spectral, or cepstral domains. It can also be extracted using auditory models. One of the time domain approaches was presented in [11]. For pitch detection, the authors proposed a so-called Yin estimator. The inspiration was the yin-yang philosophical principle of balance, which represents the authors attempt to achieve a balance between autocorrelation and cancellation, which are both implemented in the proposed algorithm. The problem of using the autocorrelation approach for pitch value estimation is that the peaks can also occur at sub-harmonics. Because of this phenomenon, it is difficult to determine which peak represents the actual fundamental frequency and which is a sub-harmonic. Yin estimation tries to solve these problems by using a difference function, which attempts to minimise the difference between the original waveform and its delayed copy. Another time domain approach is presented in [12]. The proposed approach is dealing with time domain pitch value estimation for telephone speech. Telephone speech is specific because it has reduced bandwidth and, consequently, the fundamental frequency can be very weak or even missing. In such circumstances, traditional methods based on autocorrelation cannot provide good results. To reduce the effect of narrower bandwidth, the authors propose a nonlinear filter which restores the weak or missing frequency band. After that, the combined autocorrelation function is calculated based on the original and nonlinearly processed speech. Results show 1% improvement for clean studio speech and 3% improvement for telephone speech. The experiments were performed on a Keele pitch database [13].

Pitch value can also be derived in the spectral domain, where one of the popular principles for this task is the use of tuneable filters. In [14], the author presented a method based on a narrow user-tuneable band-pass filter, which is swept across the frequency spectrum. The fundamental frequency is detected when maximum value is present on the output of the filter. The F_0 is then equal to the central frequency of the filter. The author of the paper also suggests that the difference could be detected between an evenly spaced spectrum and a richly harmonic single note. Another method using multiple comb-filter approaches was presented in [15]. The authors are investigating the problem of multiple fundamental frequencies estimation in a noisy environment. This can happen when many persons speak at the same time with the presence of background noise. Their work is done for two speakers. The pitch value of the first speaker is determined by detecting the autocorrelation of the multi-scale product (AMP) of the mixture signal. After that, a multiple comb filter is applied to filter out the dominant signal. A residual signal is obtained after the subtraction of the remaining signal from the mixture signal. Next, the AMP is applied to the residual signal to estimate the pitch value of the second speaker. Results of the proposed method show that the method is robust and effective. Experiments were performed on the Cooke database [16]. The pitch estimation algorithm, which is robust against high levels of noise, called PEFAC, was proposed by Gonzales and Brookes [17]. The algorithm is able to identify voiced frames and estimate pitch reliably, even at negative signal-to-noise ratios. The proposed principle uses nonlinear amplitude compression to reduce narrowband noise for more robust pitch estimation. Two Gaussian mixture models (GMMs) are trained for voiced speech detection and are used for voiced/unvoiced speech classification. The proposed algorithm was evaluated on a part of the TIMIT database and on the CSLU-VOICES corpus and compared with other widely used algorithms. The tests show better performance, especially for negative SNR. The authors in [18] proposed robust harmonic features for classification-based pitch estimation. The proposed pitch estimation algorithm is composed of pitch candidate generation and target pitch selection stages. Two types of spectrum are proposed for extracting pitch candidates. One is the original noisy long-term speech spectrum, and the other is the long-term sub-harmonic summation (SBH) spectrum. If the SNR is low in the part where the F_0 is present, the F_0 spectral peak could disappear. In this case, SBH serves as a complementary source for pitch candidate extraction. In the second step of the proposed algorithm, pitch candidate classification using a neural network is performed, based on multidimensional pitch-related robust harmonic features. The five proposed features are based on the energy intensity and spectrum envelope properties of the speech. Experiments were performed on the Keele database and CSTR database. Performance of the proposed algorithm was tested against five of the common pitch estimation algorithms, including SAcC, JinWang, PEFAC, RAPT, and Yin. The results show better performance than the compared algorithms across various types and levels of noise.

Another domain where methods for pitch value extraction exist is the cepstral domain. A cepstrum is a form of spectrum where the output is the Fourier transform of the logarithmic spectral magnitude of the original waveform. The author in [19] proposed a method which needed a jury of experienced listeners for pitch value estimation judgement. Cepstrum was computed digitally and then transformed on microfilm by plotter. The method was proposed in 1967, when computer use for processing was still minimal. Another cepstrum-based method for fundamental frequency estimation was presented in [20]. Pitch information is extracted using a modified cepstrum-based method, after which the cepstrum is refined using a pitch value tracking, correction, and smoothing algorithm. In the presented work, a cepstrum-based voicing detector is also discussed. Voicing decisions are made using a multi-featured voiced/unvoiced (V/UV) classification algorithm, based on statistical analysis of the zero-crossing rate, energy of short-time segments, and cepstral peaks. Experiments were performed on speech data taken from TIMIT database. Results show considerable improvement relative to the conventional cepstrum methods. The proposed algorithm also tends to be robust against additive noise.

Pitch value can also be derived using auditory models, as the author presented in [21]. They proposed a multi-channel pitch determination algorithm (PDA), which is composed of an automatic channel selection module and a pitch value extraction module that relies on the pseudo-periodic histogram for the pitch value search.

The proposed PDA outperformed the reference system (auditory modelling AMPEX) for 0 dB SNR telephone speech and car speech. The automatic selection of channels was effective on the very noisy telephone speech but performed less successfully in the car speech. Another model-based approach was proposed by Shi et al. [22]. Their approach uses Bayesian pitch tracking, which is based on the harmonic model. Good robustness against noise was achieved by using the parametric harmonic model. A fully Bayesian approach was applied to avoid overfitting of the first-order Markov chains. Results show that the proposed algorithm has good robustness against voicing state changes, as it carries past information on pitch over the unvoiced or silent regions. Experiments were performed on Keele and Parkinson's disease databases.

Amongst other things, pitch estimation is a very important feature for the speaker's gender classification, as it is one of the more distinguishable properties between male and female speakers. Information about the speaker's gender is useful for tasks like speaker clustering or demographic data collection. In work presented in [23], the author used formant and energy-based features and several different pitch-based features for speaker's gender classification on emotionally coloured speech. Some of the features used are min., max., average pitch values, interquartile pitch range. Experiments were performed on the Danish emotional speech (DES) database, Sahad emotional speech (SES) database, and German emotional speech (GES) database. A probabilistic neural network (PNN) is a feedforward neural network, which is widely used in classification, support vector machines (SVM), K-nearest neighbour (K-NN), and GMM were compared for classification performances of naive Bayes. Results show over 90% gender classification accuracy, where the SVM classifier gave the best results.

The next section will give a general view of the determination of the pitch value in the speech signal and present what needs to be done to determine the pitch value at all.

3. Generally on the pitch value determination

Determining the speaker pitch value from the captured audio signal is possible, both in time and frequency domain representation of the signal. Determination is possible only in parts of the signal that contain a voiced speech signal. Representatives of the voiced speech signal are vowels, diphthongs, and semivowels that contain much more energy than the consonants, which are also present in the speech signal. **Figure 2** shows the time and frequency domain presentation of the vowel /eh/ and the consonant /s/ of the word "seven" in the captured audio signal. There is a considerable difference in the amplitude between the vowel /eh/ and the consonant /s/ in both the time and frequency domains. The amplitude of the vowel /eh/ (**Figure 2(a)**) is about 100 times greater than the consonant amplitude /s/ (**Figure 2(c)**). **Figure 2(a)** also has a well-seen repetitive signal pattern from which it is possible to determine a pitch value, whilst, for **Figure 2(c)**, this cannot be said. In the time domain, the pitch value or fundamental frequency F_0 can be calculated as:

$$F_0 = \frac{f_{samp}}{\tau}, \tag{1}$$

where f_{samp} is the sampling frequency of the captured audio signal, and τ is the difference between the peaks. The determination of the last value is presented in **Figure 2(a)**. In this case, the value τ is (133−96) = 37 samples, whilst the sampling frequency is 8000 Hz. From this, it follows that the pitch value is equal to 216.2 Hz. The determination of the pitch value in the frequency domain is

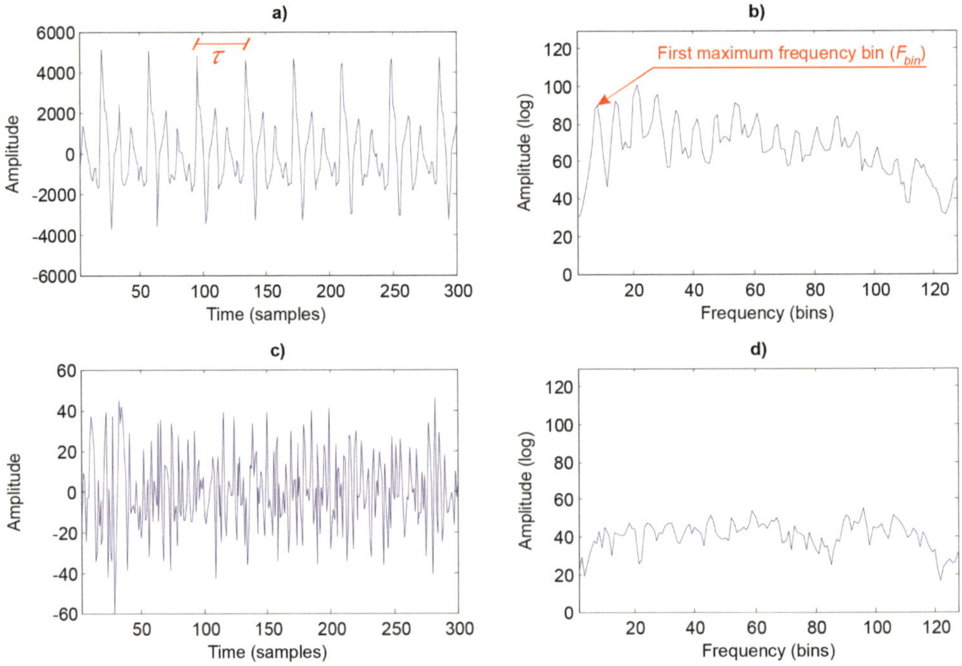

Figure 2.
Time and frequency domain presentation of the vowel /eh/ and the consonant /s/ of the word "seven" in the captured speech audio signal. (a) Voiced vowel /eh/ in time domain, (b) voiced vowel /eh/ in frequency domain, (c) unvoiced consonant /s/ in time domain and, (d) unvoiced consonant /s/ in frequency domain.

presented in **Figure 2(b)**. The pitch value can be determined by detecting the first maximum value on the frequency axis, and it is calculated as:

$$F_0 = F_{bin} \frac{f_{samp}}{2 \times F_{allbins}}, \tag{2}$$

where f_{samp} is the sampling frequency of the captured audio signal, F_{bin} is the first maximum value on the frequency axis, and $F_{allbins}$ is the number of all bins on the frequency axis. In **Figure 2(b)**, the value F_{bin} is 7 bins, the value $F_{allbins}$ is 128 bins, and the sampling frequency is 8000 Hz. So, it follows that the pitch value is 218.8 Hz. The difference between the two calculated pitch values on the same frame of the speech signal is less than 3 Hz. In the areas of the speech signal, where the consonants are located, it is not possible to determine the pitch value. Therefore, it is very important to define the boundaries of the voiced signal correctly in the whole speech signal. The voice activity detection (VAD) algorithm determines the presence of a voiced speech signal.

The VAD algorithm usually detects the presence of the entire speech signal in the captured audio signal. Such a solution is used in ASR systems to improve speech recognition accuracy. In methods for pitch value extraction, however, it is important that the VAD algorithm detects parts that contain only voiced parts of the speech signal. **Figure 3** shows the result of the voiced speech detection for the word "seven", which is obtained with the VAD algorithm.

Once the voiced parts of the speech signal are defined, then the determination of the pitch value can be made on these parts of the signal. However, another problem occurs when detecting the pitch value of a particular speaker. The pitch value of the speaker changes through pronunciation and is not constant at all times. If, as an

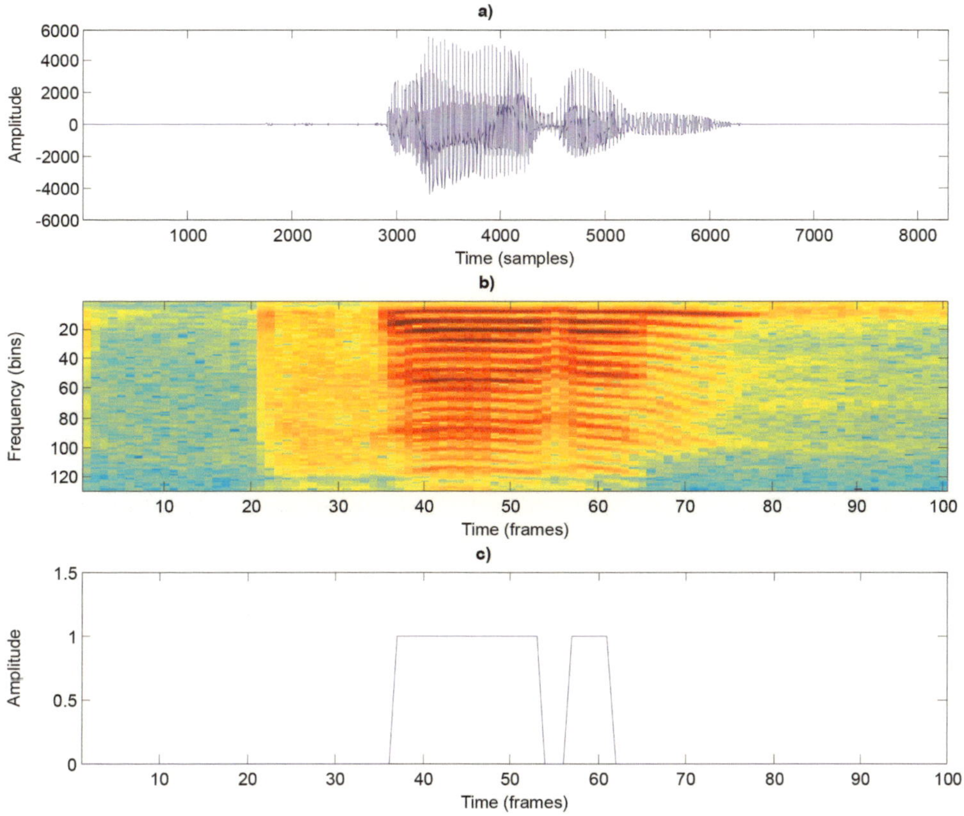

Figure 3.
VAD detection for only voiced parts of the speech signal for the word "seven": (a) audio signal in the time domain, (b) audio signal in the frequency domain, and (c) VAD decision.

example, the pronunciation of the word "seven" is taken, which is presented in **Figure 3**, the phonetic record of the word "seven" is /s eh v ah n/. In the word "seven", there are two vowels, namely /eh/ and /ah/, which represent the voiced part of the speech, in which the pitch value can be determined. For vowel /eh/, the pitch value of 216.2 Hz is determined, whilst for the vowel / ah /, the value is 222.2 Hz. As can be seen, the pitch values differ although they are very close. Because these vowels are very similar, there are no significant differences. Otherwise, there is a difference by the pronunciation of the word "zero". The phonetic record of the word "zero" is /z ih r ow/. The pitch values are determined on the vowels /ih/ and /ow/. The first vowel has a pitch value of 266.6 Hz, whilst the other has a value of 190.5 Hz. As can be seen, there are substantial differences between the calculated values although the vocals are part of one word spoken by one speaker. Such significant differences for isolated words occur but not often. However, even more significant fluctuations in pitch value detection occur in longer sentences, as the speakers, in particular, of a declarative sentence, start to speak loudly and more quietly towards the end of the sentence. This type of speech, however, contributes to a greater fluctuation of the pitch value for one speaker. Therefore, in the presented chapter, the comparative tests are made on short words, such as isolated digits.

The next section will present the process of determining the pitch value, which works well in different noise environments and also for low signal-to-noise ratios (SNRs).

4. The proposed pitch value detection procedure

As already mentioned, when determining the pitch values, it is primarily necessary to determine where the voiced speech areas are in the captured audio signal. The next subsection will present the VAD algorithm that was used to detect the voiced part of the speech for the pitch value determination process. After that, the description of the procedure will be presented which defines the pitch value in the voiced areas of the speech signal.

4.1 Voice activity detection algorithm

The voice activity detection (VAD) algorithm could have an improper effect on the results of pitch value detection. For this reason, the boundaries are determined of the beginning and end of the presence of a voiced signal in a speech signal on a clean signal only, without the presence of noise. The resulting boundaries were then also used for audio recordings with added different noisy signals with different SNR values. In order to explain the process of determining the VAD decision on the clean speech signal, **Figure 4** will be used, in which the audio signal is presented in the time domain (**Figure 4(a)**) and the frequency domain (**Figure 4(b)**) for the spoken word "four". **Figure 4(c–e)** present frame energy values and zero-crossing measure values with the corresponding threshold values.

The values of frame energy E_f are presented in **Figure 4(c)** as a blue line. The value of frame energy E_f is calculated as:

$$E_f = \frac{\sum_{i=1}^{N} s[i]^2}{N} \tag{3}$$

where $s[i]$ is a sample of an audio signal, and N the number of samples in the frame, which, in our case, is 300 samples. The energy threshold E_{th} (the red line in **Figure 4(c)**) is defined for a whole audio signal and is calculated as:

$$E_{th} = \frac{\max(E_f) + \min(E_f)}{2} \tag{4}$$

where max (E_f) is the maximum frame energy value, and min (E_f) is the minimum frame energy value in the whole audio signal.

The frame zero-crossing measure value, denoted with ZC_f, presents how many times the signal in the frame crosses the value zero or changes the sign. The zero-crossing measure value gives us additional information for the VAD decision, since it is widely known that a large zero-crossing measure value in the frame represents the frame that contains noise or frame in the audio signal which contains unvoiced speech. For example, phoneme /f/ is a consonant, which belongs to unvoiced speech. **Figure 4(d)** shows the frame zero-crossing measure value (blue line). It can be concluded from **Figure 4(d)** that the ZC_f values in the regions of unvoiced speech and noise signal are indeed large and much larger than those in the region of a voiced speech signal. The zero-crossing threshold value ZC_{th} can be set, which determines the segments of unvoiced speech and segments of voiced speech signal as:

$$ZC_{th} = \frac{\max(ZC_f) + \min(ZC_f)}{2} \tag{5}$$

where max (ZC_f) is the maximum zero-crossing measure value, and min (ZC_f) is the minimum zero-crossing measure value in the whole audio signal. The zero-crossing threshold value ZC_{th} is presented in **Figure 4(d)** as a red line.

The VAD decision is based on frame energy value E_f, frame zero-crossing measure value ZC_f, threshold energy value E_{th}, and zero-crossing threshold value ZC_{th}, as presented in Eq. (6). The proposed VAD algorithm detects voiced speech frames, since only in these frames of the speech signal, the pitch value can be determined. **Figure 4(e)** shows the VAD decision on the whole audio signal.

$$VAD_f = \begin{cases} 1; & (E_f > E_{th}) \wedge (ZC_f < ZC_{th}) \\ 0; & (E_f \le E_{th}) \wedge (ZC_f \ge ZC_{th}) \end{cases} \tag{6}$$

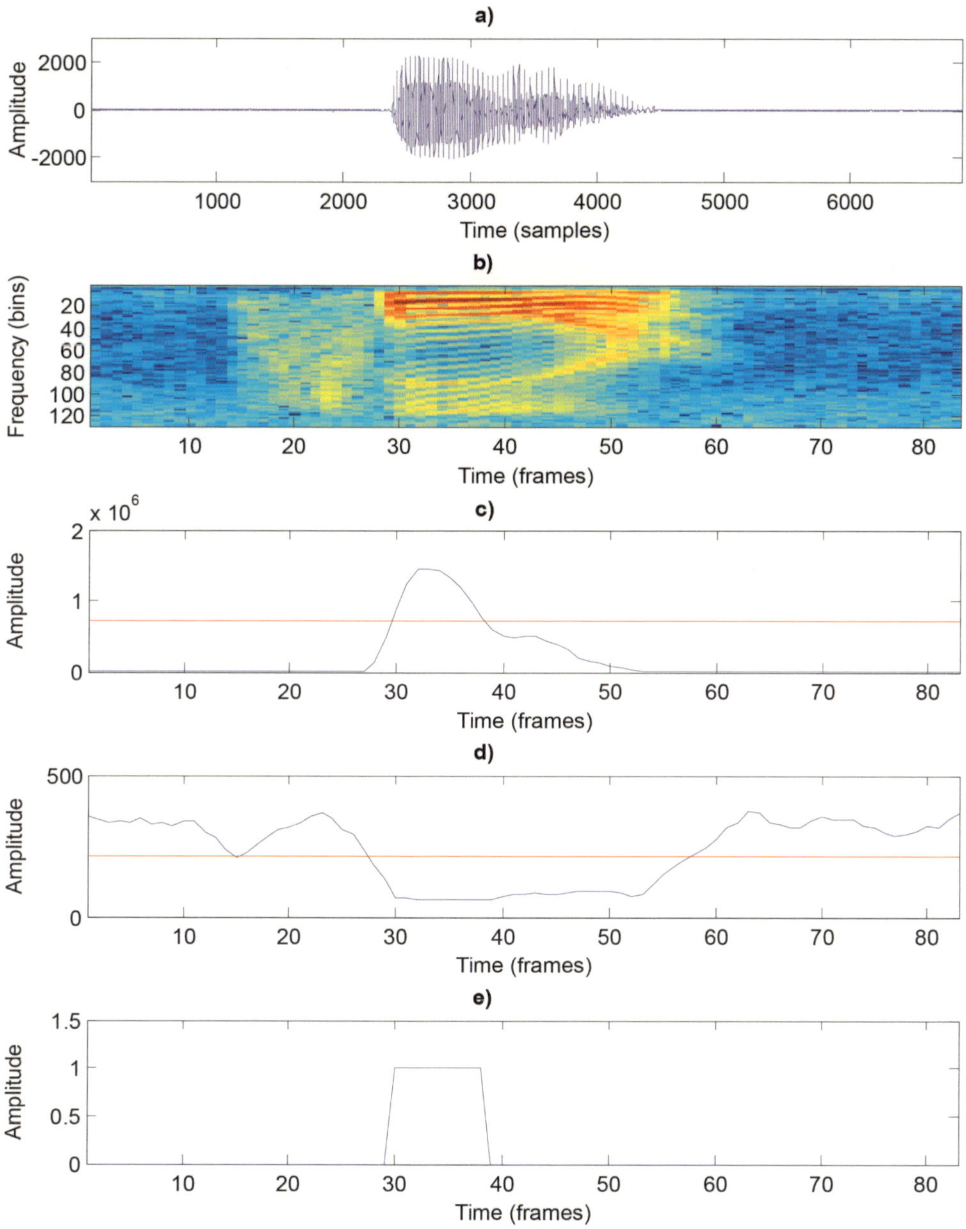

Figure 4.
VAD decision on clean speech signal for the word "four": (a) audio signal in the time domain, (b) audio signal in the frequency domain, (c) frame energy values and energy threshold value, (d) zero-crossing measure values and zero-crossing threshold value, and (e) VAD decision.

4.2 Pitch value detection

The time domain representation of the audio signal is used to determine the pitch value. The pitch value is determined in each frame which was previously detected by the VAD algorithm as a voiced part of the speech signal. To explain the process of determining the pitch value, **Figure 5** will be used, in which the part of the speech signal is presented where the word "three" is pronounced. The presentation on the vowel /iy/ will be made, which is located at the end of the pronunciation of this word.

The first step in the pitch calculation procedure is to define the highest maximum value between positive samples values and the lowest minimum value between negative samples values. The red line in **Figure 5(a)** presents the highest maximum and the lowest minimum values of the samples in the frame. The next step is to define the positive and negative peaks. Only the samples that are greater than 75% of the maximum or minimum value are used and searched for the current peak maximum or minimum. The maximums are searched in the direction of the

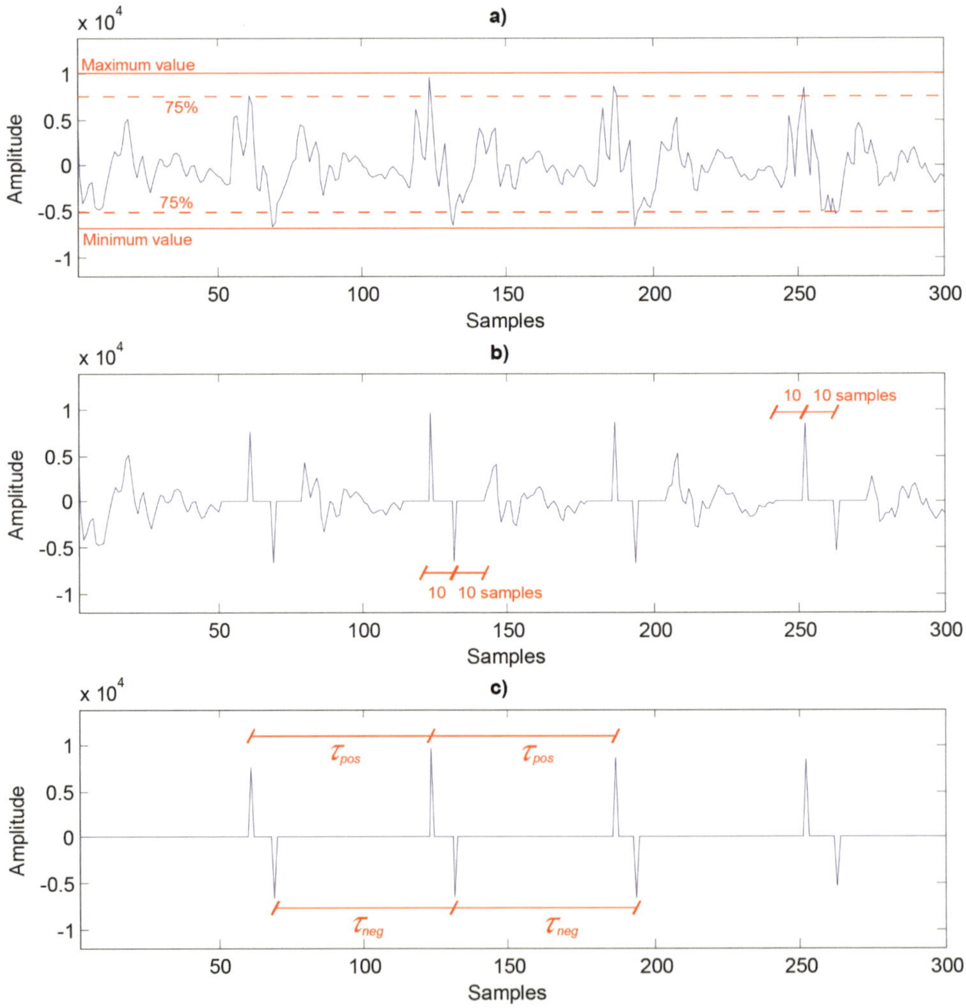

Figure 5.
The time domain representation of one frame on phoneme /ah/: (a) search for peaks in a voiced speech signal frame, (b) extraction of peaks, where 10 samples left and right around the peak are set to 0, and (c) all samples smaller than 75% of maximum and minimum values are set to 0.

highest maximum to 75% of their value. Whenever a positive or negative peak is found, the 10 samples left and right from the current positive or negative peak is set to 0. The result of this procedure is shown in **Figure 5(b)**. When all the positive and negative peaks are found, all other samples below 75% of the highest maximum or lowest minimum are set to 0. The result can be seen in **Figure 5(c)**.

The next step is to find the difference or the number of samples between the peaks. In **Figure 5(c)**, the difference is represented by the variable τ. Differences are calculated between all adjacent peaks. **Table 1** shows positive and negative peaks' positions in the presented frame and calculated differences between adjacent peaks. The difference between the last two minimum peaks is greater (look at **Figure 5(a)**); it can be seen that the minimum peak was detected incorrectly. An error occurred because the amplitude of the signal had changed slightly in that part. Just as the differences between the peaks for this represented frame can be determined, the same is done for all the frames containing the voiced speech signal. Thus, for each audio recording, a set of positive τ_{pos} and negative τ_{neg} differences is obtained between the peaks. The most commonly detected difference in each audio recording is then used to calculate the pitch value. The pitch value is determined, so that a positive pitch value F_{0pos} is obtained, with the most commonly detected positive difference between adjacent peaks and negative pitch value F_{0neg} being obtained with the most commonly detected negative difference between nearby peaks. The positive pitch value is calculated as presented in Eq. (7), and the negative pitch value as shown in Eq. (8). In Eqs. (7) and (8) the variable f_{samp} represent the sampling frequency.

$$F_{0pos} = \frac{f_{samp}}{\tau_{pos}} \tag{7}$$

$$F_{0neg} = \frac{f_{samp}}{\tau_{neg}} \tag{8}$$

If the two pitch values are the same, then the pitch value was probably determined correctly. However, if they differ, the correct pitch value is the smallest, and it is determined as:

$$F_0 = \begin{cases} F_{0pos} & F_{0pos} \leq F_{0neg} \\ F_{0neg} & F_{0pos} > F_{0neg} \end{cases} \tag{9}$$

Positive peak position	Positive difference between adjacent peaks τ_{pos}
61	
124	63
187	63
252	65
Negative peak position	Negative difference between adjacent peaks τ_{neg}
69	
132	63
194	62
263	69

Table 1.
Positive and negative peaks' positions on the vowel /iy/, which is presented graphically in **Figure 5**.

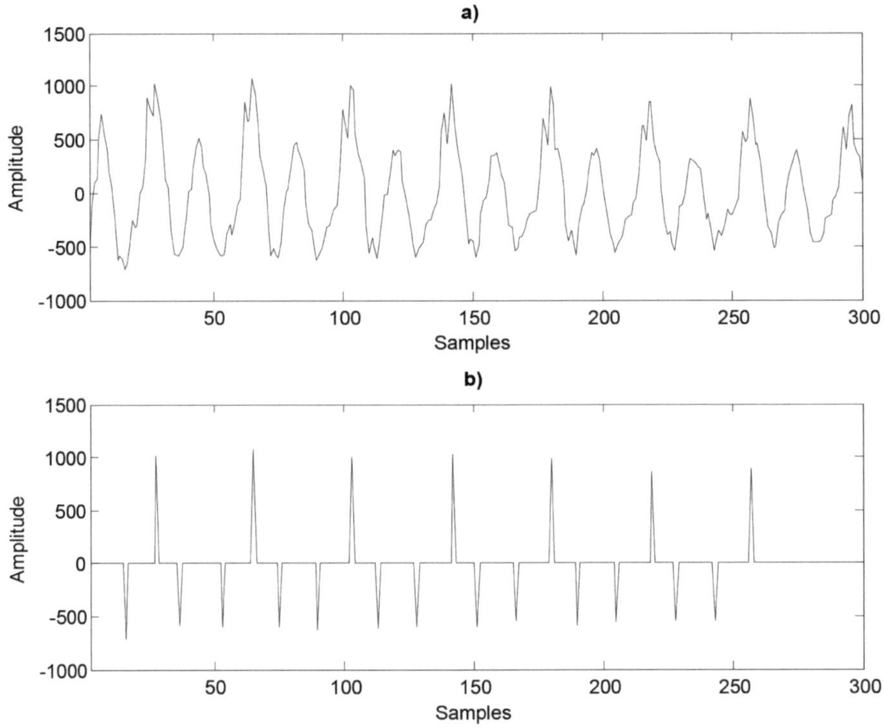

Figure 6.
Problems that may occur in peak detection: (a) time domain representation of vowel /ih/ in word "six" and (b) detected peaks' positions.

Why this decision is made can be presented in the example given in **Figure 6**. **Figure 6(b)** shows that the negative peaks are determined incorrectly. However, for positive peaks, it is evident that they are correctly defined. From this, it follows that the differences between the positive peaks are more significant than those amongst the negative peaks. To select the difference between positive peaks, consequently, the smaller pitch value is chosen.

5. Experimental design and results

In this section, the speech database is presented which was used for experiments on determining the pitch value in different noisy environments. Since the speech database used does not have reference pitch values, Subsection 5.2 will show how the reference pitch value is determined for the individual recording in the speech database. Finally, the results will be presented of the experiments on pitch value detection and gender classification.

5.1 Aurora 2 speech database

The experiments were carried out using the Aurora 2 speech database [24], which is designed to evaluate the performance of speech recognition algorithms under noisy conditions. In this chapter, the comparative tests were made only on short words, which are, in this case, isolated digits. Tests on isolated digits were chosen because, on short speech segments, the pitch value does not fluctuate so much. The speech material from the test set of the Aurora 2 speech database was used for the experiments on pitch value detection. Three different test sets were defined for the testing. Four subsets with 298, 279, 283, and 284 utterances were obtained by splitting 1144 utterances from 52 male and 52 female speakers. The recordings of all speakers were present in each subset. Individual noise signals at SNRs of 20 dB, 15 dB, 10 dB, 5 dB, 0 dB, and − 5 dB were added, and the clean case without added noise was taken as the seventh condition.

The first test set is called test set A. In this test set, four noises: a suburban train, babble, a car, and an exhibition hall were added to the four subsets. The second test set is called test set B. This test was created in the same way as test set A, the only difference is that four different noises were used, which are a restaurant, a street, an airport, and a train station. The third test set is called test set C, and it contained only the first two of four subsets, with 298 and 279 utterances. Here, speech and noise are filtered using the Motorola Integrated Radio Systems (MIRS) characteristic [25], before being added to the SNRs of 20 dB, 15 dB, 10 dB, 5 dB, 0 dB and − 5 dB. The MIRS filter represents a frequency characteristic that simulates the behaviour of a telecommunication terminal, which meets the official requirements for the terminal input frequency response as specified, e.g., in the European Telecommunications Standards Institute - Special Mobile Group (ETSI-SMG) technical specification [25]. The suburban train and street were used as added noise signals. The purpose of this set was to show the influence on pitch value when a different frequency characteristic is present in the speech signal.

Both parts, training and test material of the Aurora 2 speech database, were used for the experiments on gender classification. The gender classification experiments used the same test material as the experiments on pitch value detection. As mentioned earlier, there are 1144 audio recordings in the test set. These audio recordings are divided into 570 recordings containing male and 574 recordings containing female speakers. Most of the gender classification tests were based on GMMs. The training of GMMs requires training material, which was taken from the training part of the Aurora 2 speech base. The concept of the Aurora 2 speech database experiments includes two training modes, which are defined as training on clean data only and as training on clean and noisy (multi-condition) data. From the Aurora 2 speech database, 8440 utterances were chosen for training on clean data, which contained the recordings with 4220 male and 4220 female speakers. The same 8440 recordings were used for multi-condition training. They were divided into 20 subsets, each of which included 422 utterances. The 20 subgroups represented four different noise scenarios (a suburban train, babble, a car, and an exhibition hall) at five different SNRs.

5.2 Definition of the reference pitch values

The Aurora 2 speech database does not provide information about the pitch value in each audio recording. Therefore, the reference values on 1144 audio recordings were determined manually using graphical representations of audio recordings in the time domain. The area of voiced speech in the audio recording was determined using the VAD algorithm presented in Subsection 4.1. The reference values were determined on isolated digits, which were spoken in American English. **Table 2** lists isolated digits with phonetic transcription. Phonemes written in bold represent the vowels on which the determination of the reference pitch values was made. In determining the reference pitch value, the difference between the two peaks was defined, which, as much as possible, have similar amplitude. In **Figure 4(a)**, this would be around 2800 samples. To determine the reference pitch value for a given audio recording Eq. (1) is used. The sampling frequency of audio recordings in the Aurora 2 speech database is 8 kHz.

Word (isolated digit)	Phonetic transcription
one	w **ah** n
two	t **uw**
three	th r **iy**
four	f **ao** r
five	f **ay** v
six	s **ih** k s
seven	s **eh** v ah n
eight	**ey** t
nine	n **ay** n
zero	z **ih** r ow
oh	**ow**

Table 2.
The lists of isolated digits with phonetic transcription.

5.3 Results of pitch value determination

All the presented results in this subsection were achieved with automatic pitch value determination for each audio recording. The automatic pitch value determination was based on the procedure given in Subsection 4.2. The area of the voiced speech signal was determined on a clean speech signal with the VAD algorithm presented in Subsection 4.1. The results of the VAD algorithm were used in all recordings, including those to which different noises were added at different SNR values. The present work did not use the VAD algorithm on noisy audio recordings because it is necessary to present how well the pitch value determination algorithm works, even in noisy environments. If the VAD algorithm is also used on noisy audio recordings, the VAD algorithm could have an overwhelming effect on the pitch value determination results.

Table 3 gives the results of the absolute deviation of the automatically obtained pitch value in a positive or negative direction concerning the reference pitch value.

Test set	Noise/[dB]	Clean	SNR 20	SNR 15	SNR 10	SNR 5	SNR 0	SNR −5	Average
A	Subway	1.41	2.51	2.60	3.39	4.54	9.96	24.12	6.93
	Babble	1.46	2.60	3.09	3.68	6.12	10.87	32.17	8.57
	Car	1.56	3.18	3.14	4.09	5.42	14.95	30.57	8.99
	Exhib.	1.83	2.76	3.46	3.71	4.59	10.96	24.86	7.45
B	Rest.	1.41	2.29	2.90	3.71	5.90	11.56	25.80	7.65
	Street	1.46	2.61	2.95	3.76	4.99	14.63	25.60	8.00
	Airport	1.56	2.28	2.73	3.09	6.10	11.17	24.05	7.28
	Train	1.83	2.65	3.40	3.77	7.36	11.53	25.40	7.99
C	Subway	3.49	3.64	3.57	4.86	5.56	14.96	36.78	10.41
	Street	3.73	4.10	4.27	4.53	8.07	15.52	33.88	10.58
Overall		1.97	2.86	3.21	3.86	5.86	12.61	12.61	8.39

Table 3.
Pitch value deviation results F_{odev} in [Hz] in the positive or negative direction for the individual test set according to the reference pitch value.

For each audio recording, the F_0 value was determined automatically and compared with the reference pitch value F_{0ref}. The result F_{0dev} is a deviation from the reference value presented in Hz and was calculated as:

$$F_{0dev} = \left|F_0 - F_{0ref}\right| \tag{10}$$

The test set A in the subway set has 298 audio recordings for each SNR noise level. The presented result is the average error value in Hz over all of the 298 audio recordings in each noise level. The same results were obtained for all remaining noise sets.

Tables 4–7 give the results obtained as a percentage. **Table 4** specifies how many percentages of audio recordings have the same value as the automatically obtained pitch value relative to the reference pitch value. **Table 5** shows how many percentages of audio recordings had an error between 1 and 10 Hz, corresponding

Test set	Noise/[dB]	Clean	SNR 20	SNR 15	SNR 10	SNR 5	SNR 0	SNR −5	Average
A	Subway	63.09	54.70	51.34	50.00	42.28	34.90	16.11	44.63
	Babble	72.04	60.22	56.99	50.18	41.94	32.97	16.85	47.13
	Car	63.96	56.89	53.36	49.47	43.11	28.27	16.96	44.57
	Exhib.	63.73	54.23	45.07	45.42	35.21	29.58	17.61	41.55
B	Rest.	63.09	55.37	53.69	51.01	44.63	31.88	20.13	45.69
	Street	72.04	56.63	56.99	45.88	41.22	27.96	16.49	45.32
	Airport	63.96	59.01	52.65	54.06	41.34	40.64	24.03	47.96
	Train.	63.73	54.58	50.00	45.07	41.55	33.10	19.37	43.91
C	Subway	46.31	46.31	44.63	40.27	35.23	26.17	9.40	35.47
	Street	44.44	44.44	44.09	44.44	32.62	26.52	13.98	35.79
Overall		61.68	54.30	50.94	47.66	40.00	31.26	17.10	43.27

Table 4.
Percentage of the pitch values for the individual test set that matched the reference pitch value fully.

Test set	Noise/[dB]	Clean	SNR 20	SNR 15	SNR 10	SNR 5	SNR 0	SNR −5	Average
A	Subway	36.24	39.60	42.62	39.93	45.64	43.96	41.28	41.32
	Babble	27.24	36.56	35.13	41.94	43.73	44.44	31.54	37.22
	Car	35.34	36.75	39.93	40.64	42.40	43.11	33.22	38.77
	Exhib.	35.56	41.55	48.94	45.42	53.17	48.24	37.68	44.37
B	Rest.	36.24	40.60	39.60	40.60	42.28	44.97	32.55	39.55
	Street	27.24	36.92	36.92	44.44	48.39	44.09	42.29	40.04
	Airport	35.34	37.10	41.70	37.10	42.76	38.87	28.27	37.30
	Train.	35.56	42.25	42.96	46.13	43.31	42.96	38.03	41.60
C	Subway	45.64	43.96	46.98	47.99	49.66	43.96	30.87	44.15
	Street	44.09	43.37	46.24	42.65	46.59	40.50	27.60	41.58
Overall		35.85	39.87	42.10	42.68	45.79	43.51	34.33	40.59

Table 5.
Percentage of the pitch values for the individual test set that the error of the pitch value was between 1 and 10 Hz according to the reference pitch value.

Test set	Noise/[dB]	Clean	SNR 20	SNR 15	SNR 10	SNR 5	SNR 0	SNR −5	Average
A	Subway	0.67	4.36	5.03	7.38	8.39	8.39	12.75	6.71
	Babble	0.00	1.43	4.66	4.66	7.89	11.11	11.11	5.84
	Car	0.00	3.89	4.59	5.65	7.77	9.89	12.01	6.26
	Exhib.	0.00	2.82	3.17	7.04	8.10	10.92	15.14	6.74
B	Rest.	0.67	3.02	5.37	5.37	8.05	10.74	14.77	6.86
	Street	0.00	4.30	3.58	6.45	6.45	9.68	11.83	6.04
	Airport	0.00	1.41	3.53	6.71	9.19	7.07	16.25	6.31
	Train.	0.00	1.76	4.93	5.99	7.04	10.21	8.80	5.53
C	Subway	5.03	6.71	5.70	8.05	10.40	10.74	15.10	8.82
	Street	8.60	8.60	5.73	8.60	10.75	12.90	13.62	9.83
Overall		1.50	3.85	4.65	6.61	8.43	10.17	13.18	6.91

Table 6.
Percentage of the pitch values for the individual test set that the error of the pitch value was between 11 and 20 Hz according to the reference pitch value.

to the reference pitch value. **Tables 6** and 7 give results similar to **Table 5**, but **Table 6** represents the percentage of audio recordings with errors between 11 and 20 Hz, and **Table 7** represents the percentage of audio recordings with errors above 21 Hz.

5.4 Results on gender classification

The gender classification experiments presented in this chapter will show how important the correct pitch value detection is for the gender classification. The tests were performed on the Aurora 2 voice database, and as presented in Subsection 5.1, 1144 audio recordings were available, of which 570 were with a male speaker and 574 with a female speaker. Seven experiments were performed. The first six tests

Test set	Noise/[dB]	Clean	SNR 20	SNR 15	SNR 10	SNR 5	SNR 0	SNR −5	Average
A	Subway	0.00	1.34	1.01	2.68	3.69	12.75	29.87	7.33
	Babble	0.72	1.79	3.23	3.23	6.45	11.47	40.50	9.63
	Car	0.71	2.47	2.12	4.24	6.71	18.73	37.81	10.40
	Exhib.	0.70	1.41	2.82	2.11	3.52	11.27	29.58	7.34
B	Rest.	0.00	1.01	1.34	3.02	5.03	12.42	32.55	7.91
	Street	0.72	2.15	2.51	3.23	3.94	18.28	29.39	8.60
	Airport	0.71	2.47	2.12	2.12	6.71	13.43	31.45	8.43
	Train.	0.70	1.41	2.11	2.82	8.10	13.73	33.80	8.95
C	Subway	3.02	3.02	2.68	3.69	4.70	19.13	44.63	11.55
	Street	2.87	3.58	3.94	4.30	10.04	20.07	44.80	12.80
Overall		1.01	2.06	2.38	3.15	5.87	15.14	35.49	9.30

Table 7.
Percentage of the pitch values for the individual test set that the error of the pitch value was greater than 21 Hz according to the reference pitch value.

used GMMs, and in the seventh test, the gender classification was determined based on the pitch value. Two separate models were trained for gender classification, one for the male speaker and one for the female speaker. GMMs were trained using the procedures described in [26]. The training of the GMMs was done on the training material of the Aurora 2 speech base, which contains two training modes (clean and multi-condition). The clean training material was used for the first three of the six GMMs training processes. The results of these experiments are presented in **Figure 7**. For the other three, the multi-condition content was used, the results of which are shown in **Figure 8**. The results in both figures are given as accuracy (Acc) in the percentage of the correct speaker gender classification, and it was calculated as:

$$Acc = \frac{H}{N} \cdot 100[\%] \tag{11}$$

where H is the sum of all correct classifications for male and female, divided by the number of all classifications N, which is 2865 for all noisy conditions.

The MFCC_E_D_A features were used for the first and fourth tests, where the entire audio recording was used for training without using segmentation. The MFCC_E_D_A features consisted of 12 Mel frequency cepstral coefficients C1-C12, logarithmic energy, and the first and second derivatives of those coefficients. The determination of the MFCC_E features is described in [27]. The procedure of calculating the first and second derivatives is described in [26]. MFCC_E_D_A features were also used for the second and fifth tests, but, in this case, segmentation was used, based on the VAD algorithm presented in Subsection 4.1. In this case, only parts of the audio recordings that contain only parts of the voiced speech signal from the audio recordings were used for training. For the third and sixth tests, an additional feature was used, namely the pitch value determined in each frame. In this case, the 12th coefficient C12 was replaced by the pitch value. So, for these two tests, the MFCC_Pitch_E_D_A features were used to train the GMM models. In this case, segmentation was also used, so that only parts of audio recordings that contained only voiced speech signals were used for training. The last, seventh test, however, was performed based on determining the pitch value F_0 for each audio recording. The results of this test are given as the last set of columns in both **Figures 7** and **8**. The pitch value limit was set at 155 Hz, so that the speaker's gender classification was defined as:

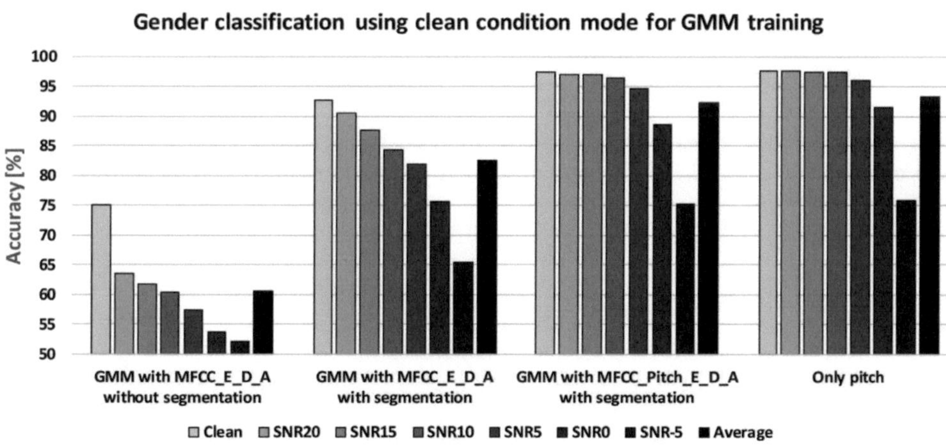

Figure 7.
Gender classification using the clean condition training mode of the Aurora 2 speech database.

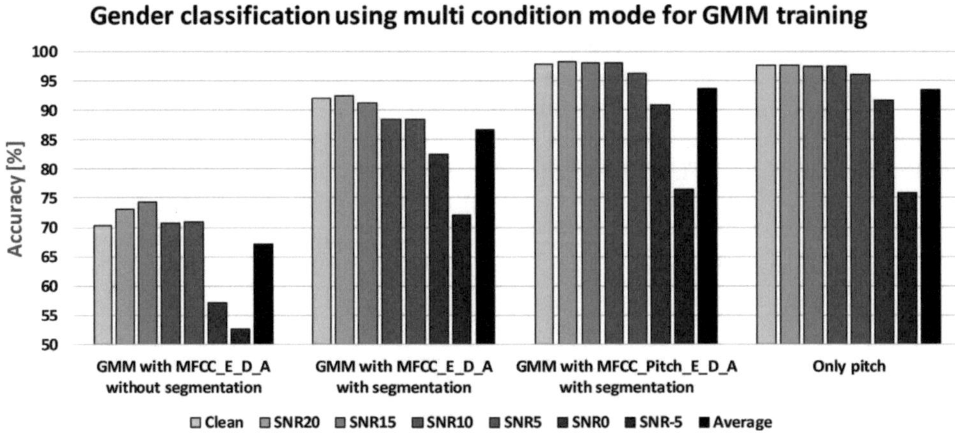

Figure 8.
Gender classification using the multi-condition training mode of the Aurora 2 speech database.

Gender	Minimum	Maximum	Average
Male	78.43 Hz	170.21 Hz	120.86 Hz
Female	131.14 Hz	275.86 Hz	206.16 Hz

Table 8.
The manual determination of the pitch values on the audio recordings obtained from the Aurora 2 speech database.

$$Gender = \begin{cases} Male & F_0 \leq 155 \\ Female & F_0 > 155 \end{cases} \quad (12)$$

The pitch value limit determination was based on the manual determination of the pitch values on the audio recordings obtained from the Aurora 2 speech database. **Table 8** gives the minimum, maximum, and average values for the pitch value for the male and female speakers, which were obtained from the manual determination of the pitch values on the audio recordings of the Aurora 2 speech database. The limits for the pitch values for a male speaker are between 78 and 171 Hz, whilst the limits for the pitch value for the female speaker are between 131 and 276 Hz. As can be seen, there is some overlapping of the pitch values. The 155 Hz pitch value limit is based on an analysis of the number of errors that could occur if the pitch for the male speaker is above 155 Hz and the pitch value for the female speaker below 155 Hz. The analysis value was also performed on the remaining pitch values between 131 Hz and 171 Hz, but the proposed pitch value limit produced the smallest number of errors.

6. Discussion

The results presented in this chapter show that the proposed automatic pitch value determination algorithm works well. For more than half of the audio recordings with the SNR higher or equal to 15 dB, the determined pitch value compared with the reference pitch value was correct (see **Table 4** for overall value). Interesting is that, on average, 17.10% full match of pitch value was achieved for audio recordings with the SNR value −5 dB. If it is taken into account that pitch differences up to 10 Hz are still an acceptable error, then, from **Tables 4** and **5**, it can be concluded that, on average, in all noisy environments with different SNR values, the algorithm can correctly detect 83.86% of the pitch values for all audio recordings.

It is evident from **Tables 6** and 7 that, even in the case of a clean signal, there are errors greater than 11 and 21 Hz compared with reference pitch value. This is due mainly to the problem described in Section 3, since the pitch value can change during the pronunciation of some words, especially if there are several different vowels in the word. At the beginning of the word, one pitch value is determined, whilst another value can be detected at the end of the word. As described in Subsection 5.2, the reference pitch value is determined on one vowel. If the word contains multiple vowels, various pitch values can be determined. In our case, the proposed pitch determination algorithm selected the pitch value that was the most often determined from the differences between the detected peaks of the voiced speech signal.

For a clean signal, the deviation of the pitch value was, on average, below 2 Hz (see **Table 3**). As can be seen from the same table, the maximum deviation of the average values in Hz was made by the test set C. In this test set, the audio recordings were filtered with an MIRS filter, which simulates the behaviour of the telecommunications terminal. The frequency response of the MIRS filter is presented in [25]. If the values of test set A for the subway noise set and test set C for the subway noise set for a clean signal are compared, the average deviation value from the reference value is 2.08 Hz when the audio recordings were filtered with the MIRS filter. In **Figure 9**, an example of the word "five" before (test set A) and after (test set C) is filtered with the MIRS filter. As can be seen, the amplitude of the speech signal is about one-third smaller after the filter was used.

Figure 10 shows the process of determining the peaks in a voiced speech signal. As can be seen, there are errors in peak detection, especially when the signal was filtered with the MIRS filter (test set C).

Based on the good determination of the pitch value, the obtained results can be used in the gender classification. As can be seen from the results presented

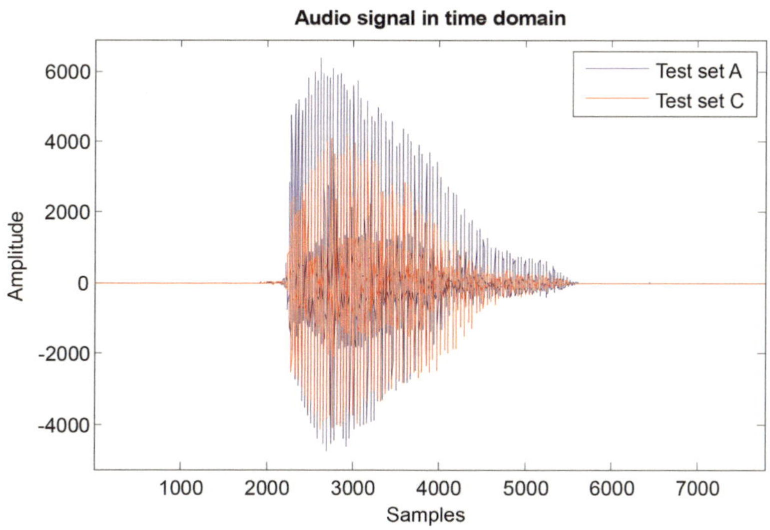

Figure 9.
The audio signal in the time domain of the word "five" before (blue line) and after (red line) it was filtered with the MIRS filter.

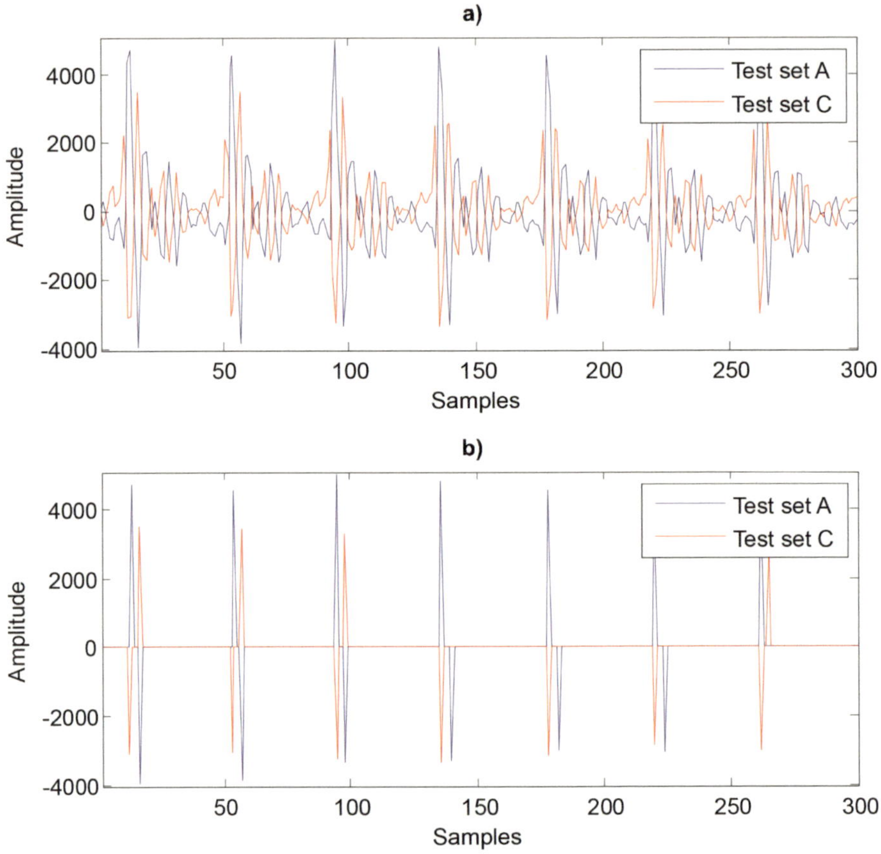

Figure 10.
The signals before (blue line) and after (red line) were filtered with the MIRS filter: (a) time domain representation of vowel /ay/ in the word "five" and (b) detected peaks positions.

in **Figures 7** and **8**, well-defined pitch values contributed significantly to the accuracy of the speaker's gender classification in both training modes (see the third set of columns in both figures). In both Figures, the fourth set of column presents results, where only the pitch value is used to classify speaker gender. The results show clearly that, even at low signal-to-noise ratios (SNR = 5 dB), the pitch value determination allowed good classification of the speaker's gender. Speaker gender classification accuracy is above 96% for SNR5. In this case, the performance was better than the performance using GMM models. However, when using GMMs, the speaker's gender classification results can be better if more training material is used. If only the pitch value is used for classification, using a different speech database will likely require a new pitch limit value to be defined. The results show, however, that the pitch value used as an additional coefficient for the features contributed greatly to the accuracy of the speaker's gender classification.

However, once a useful speaker's gender classification is made, then this can be used in intelligent environments, where the performance of natural language processing can be improved.

7. Conclusion

An effective determination of the pitch values, which works well in various noise environments, is presented in this chapter. At the beginning of this chapter, an overview is made of the pitch values used in the technologies of natural language processing. After that, the general procedures are presented for determining the pitch value in the time and frequency domains. The main part of this chapter is the presentation of the proposed procedure for determining the pitch values. The experiments were carried out on a part of the Aurora 2 speech database. Only isolated digits were used in the tests. Isolated digits represent short words on which the pitch value can be determined without major changes during the speech pronunciation. As presented in this chapter, this may also happen in short words and even more often with longer sentences. The results showed that automatically determined pitch values for all noisy environments deviated, on average, by 8.39 Hz compared with the reference pitch values.

A well-defined pitch value allows a functional speaker's gender classification. The pitch value determination procedure presented in this chapter provides a good speaker's gender classification, even at low signal-to-noise ratios. Thus, when the automatically determined pitch value is used, the speaker's gender classification performance at SNR 0 dB is higher than 91%. A speaker's gender classification can be then used further in the processes of natural language processing.

Author details

Damjan Vlaj*, Andrej Žgank and Marko Kos
Faculty of Electrical Engineering and Computer Science, University of Maribor, Maribor, Slovenia

*Address all correspondence to: damjan.vlaj@um.si

References

[1] Pieraccini R, Lubensky D. Spoken language communication with machines: The long and winding road from research to business. In: International Conference on Industrial, Engineering and Other Applications of Applied Intelligent Systems. Berlin, Heidelberg: Springer; 2005. pp. 6-15

[2] Côté N, Berger J. Speech communication. In: Möller S, Raake A, editors. Quality of Experience. T-Labs Series in Telecommunication Services. Cham: Springer; 2014

[3] Vacher M, Istrate D, Portet F, Joubert T, Chevalier T, Smidtas S, et al. The sweet-home project: Audio technology in smart homes to improve well-being and reliance. In: 2011 Annual International Conference of the IEEE Engineering in Medicine and Biology Society. 2011. pp. 5291-5294

[4] Brdiczka O, Langet M, Maisonnasse J, Crowley JL. Detecting human behavior models from multimodal observation in a smart home. IEEE Transactions on Automation Science and Engineering. 2008;**6**(4):588-597

[5] Besacier L, Barnard E, Karpov A, Schultz T. Automatic speech recognition for under-resourced languages: A survey. Speech Communication. 2014;**56**:85-100

[6] Giannakakis G, Grigoriadis D, Giannakaki K, Simantiraki O, Roniotis A, Tsiknakis M. Review on psychological stress detection using biosignals. IEEE Transactions on Affective Computing. 2019. DOI: 10.1109/TAFFC.2019.2927337

[7] Wanner L, André E, Blat J, Dasiopoulou S, Farrùs M, Fraga T, et al. Kristina: A knowledge-based virtual conversation agent. In: International Conference on Practical Applications of Agents and Multi-Agent Systems. Cham: Springer; 2017. pp. 284-295

[8] Mary L. Significance of prosody for speaker, language, emotion, and speech recognition. In: Extraction of Prosody for Automatic Speaker, Language, Emotion and Speech Recognition. Cham: Springer; 2019. pp. 1-22

[9] Drugman T, Huybrechts G, Klimkov V, Moinet A. Traditional machine learning for pitch detection. IEEE Signal Processing Letters. 2018; **25**(11):1745-1749. DOI: 10.1109/LSP.2018.2874155

[10] Gerhard D, Pitch extraction and fundamental frequency: History and current techniques, Technical Report TR-CS 2003-06; 2003

[11] de Cheveigne A, Kawahara H. Yin, a fundamental frequency estimator for speech and music. Journal of the Acoustical Society of America. 2002; **111**(4):1917-1930

[12] Chang L, Xu J, Tang K, Cui H. A new robust pitch determination algorithm for telephone speech. In: 2012 International Symposium on Information Theory and its Applications. Honolulu, HI; 2012. pp. 789-791

[13] Plante F, Meyer G, Ainsworth WA. A pitch extraction reference database. In: EUROSPEECH'95. Madrid; 1995. pp. 837-840

[14] Lane JE. Pitch detection using a tunable IIR filter. Computer Music Journal. 1990;**14**(3):46-59

[15] Zeremdini J, Anouar M, Messaoud B, Bouzid A. Multiple comb filters and autocorrelation of the multi-scale product for multi-pitch estimation. Applied Acoustics. 2017;**120**:45-53. DOI: 10.1016/j.apacoust.2017.01.013

[16] Cooke M, Barker J. An audio-visual corpus for speech perception and automatic speech recognition. Journal of Acoustic Society of America. 2006; **120**(5):2421-2424

[17] Gonzalez S, Brookes M. PEFAC—A pitch estimation algorithm robust to high levels of noise. IEEE/ACM Transactions on Audio, Speech, and Language Processing. 2014;**22**(2): 518-530

[18] Wang D, Yu C, Hansen JHL. Robust harmonic features for classification-based pitch estimation. IEEE/ACM Transactions on Audio, Speech, and Language Processing. 2017;**25**(5): 952-964

[19] Noll A. Cepstrum pitch determination. Journal of the Acoustical Society of America. 1967;**41**(2):293-309

[20] Ahmadi S, Spanias AS. Cepstrum-based pitch detection using a new statistical V/UV classification algorithm. IEEE Transactions on Speech and Audio Processing. 1999;**7**(3):333-338

[21] van Immerseel L, Martens J. Pitch and voiced/unvoiced determination with an auditory model. Journal of the Acoustical Society of America. 1992; **91**(6):3511-3526

[22] Shi L, Nielsen JK, Jensen JR, Little MA, Christensen MG. Robust Bayesian pitch tracking based on the harmonic model. IEEE/ACM Transactions on Audio, Speech, and Language Processing. 2019;**27**(11): 1737-1751

[23] Sedaaghi MH. Gender classification in emotional speech. In: Mihelic F, Zibert J, editors. Speech Recognition. Rijeka: IntechOpen; 2008. DOI: 10.5772/6385

[24] Hirsch HG, Pearce D. The Aurora experimental framework for the performance evaluation of speech recognition systems under noisy conditions. In: Proceedings of ISCA Tutorial and Research Workshop (ITRW) on ASR. Paris, France; 2000

[25] ETSI-SMG Technical Specification, European digital cellular telecommunication system (phase 1)—Transmission planning aspects for the speech service in GSM PLMN system, ETSI-SMG technical specification GSM03.50, Version 3.4.0. Valbonne, France; 1994

[26] Young S, Evermann G, Gales M, Hain T, Kershaw D, Liu XA, et al. The HTK book, Version 3.4. Cambridge University Engineering Department; 2006

[27] ETSI Standard, Speech processing, transmission and quality aspects (STQ), distributed speech recognition, front-end feature extraction algorithm, compression algorithm, ETSI Standard ES 201 108 v1.1. Valbonne, France; 2000

Section 2

Data Mining and Its Applications

Chapter 4

Classification Problem in Imbalanced Datasets

Aouatef Mahani and Ahmed Riad Baba Ali

Abstract

Classification is a data mining task. It aims to extract knowledge from large datasets. There are two kinds of classification. The first one is known as complete classification, and it is applied to balanced datasets. However, when it is applied to imbalanced ones, it is called partial classification or a problem of classification in imbalanced datasets, which is a fundamental problem in machine learning, and it has received much attention. Considering the importance of this issue, a large amount of techniques have been proposed trying to address this problem. These proposals can be divided into three levels: the algorithm level, the data level, and the hybrid level. In this chapter, we will present the classification problem in imbalanced datasets, its domains of application, its appropriate measures of performances, and its approaches and techniques.

Keywords: classification, imbalanced datasets, sampling, data mining, classifier

1. Introduction

Classification is the most popular task of data mining. It consists of assigning to each instance a class chosen from a set of predefined classes, according to the value of certain predictive attributes [1]. Its problem is to correctly classify an instance with indeterminate class. This classification can be done by several methods that are divided into two categories. The first category is based on the use of a model or a classifier such as decision trees and classification rules. However, the second category is based on the internal functioning of the learning algorithm such as neural networks [2] and support vector machines (SVMs). All these methods use large datasets to extract knowledge.

The used datasets are organized in the form of tables. The tables' columns are called the attributes, and they represent the characteristics of the dataset. Traditionally, the last attribute is called a class attribute. The tables' rows represent the data, and they are called instances. The number of instances varies from one class to another. So, the number of instances of one class is larger than that of the second class in some existing datasets. Therefore, datasets are divided into two categories: balanced and imbalanced datasets. In the latter, instances are divided into two sets: majority instances which are the most frequent and minority instances which are the less frequent.

Rule-based classification algorithms have a bias toward majority classes [3]. They tend to discover the rules with high values of accuracy and coverage. These rules are usually specific to majority instances, whereas specific rules that predict

 IntechOpen

minority instances are usually ignored or treated as noise. Consequently, minority instances are often misclassified. Generally, because most classifiers are designed to minimize the global error rate [4], many problems occur. First, they perform poorly on imbalanced datasets, and they either produce general rules or very specific ones. In the first case, the classifier has a bias toward majority instances, and it ignores the minority ones. In the second case, the classifiers tend to overfit the training data which provokes poor classification accuracy on unseen data. Next, the cost of misclassifying a minority instance is usually more expensive than misclassifying a majority one [5, 6]. Finally, in many applications misclassifying a rare event can result in more serious problems than a common event [7]. For example, in case of cancerous cell detection in medical diagnosis, misclassifying non-cancerous cells may lead to some additional clinical tests, but misclassifying cancerous cells leads to very serious health risks.

The class imbalance problem is a fundamental problem in machine learning, and it has received much attention [8–14]. This problem is known as partial classification [15], nugget discovery [16], classification problem with imbalanced datasets [17], or datasets with rare classes [18]. Considering the importance of this issue, a large amount of techniques have been developed trying to address this problem. These proposals can be divided into three groups which depend on how they deal with class imbalance. First, the algorithm-level approaches can either propose specific algorithms or modify the existing ones. Second, the data-level techniques introduce an additional processing step to decrease the effect of skewed class distribution such as undersampling and oversampling methods. Finally, the hybrid-level methods combine algorithm level and data level such as boosting and cost-sensitive learning.

This chapter is organized as follows. Section 2 presents the classification problem in imbalanced datasets. In Section 3, we present some domains in which the datasets appear. In Section 4, we present the evaluation metrics used in classification problem in imbalanced datasets. In Section 5, we detail the different approaches and techniques used to handle classification in imbalanced datasets. Finally, in Section 6, we make our concluding remarks.

2. Presentation of the classification problem

In the binary imbalanced datasets, the number of instances of one class is higher than that of the second class. Consequently, the first class is known as majority class and the second class as minority one. Therefore, this dataset contains two kinds of instances: majority and minority.

The distribution of instances in imbalanced binary datasets is measured by the imbalanced ratio (IR) [19] which is defined in Eq. (1):

$$IR = \frac{Number\ of\ majority\ instances}{Number\ of\ minority\ instances} \tag{1}$$

According to the value of IR, the imbalanced datasets are divided into three classes [20]: datasets with low imbalance (IR is between 1.5 and 3), datasets with medium imbalance (IR is between 3 and 9), and datasets with high imbalance (IR is higher than 9).

3. Application domains

The imbalanced datasets appear in the following several domains.

3.1 Risk management

Every year, the telecommunication industry suffers billions of dollars in unrecoverable debts. Therefore, uncollectible control is a major problem in the industry. One solution is to use large amounts of historical data to build models that are used to assess risk for each customer client or for each transaction to support risk management that reduces the level of unrecoverable debt. However, in a dataset, nonpayment of customers includes a few percent of the population [21].

3.2 Medical diagnosis

Clinical datasets store large amounts of patient information. Data mining technique is applied on these datasets to uncover the relationships and trends between clinical and pathological data. It aims to understand the evolution and characteristics of certain diseases. However, in these datasets, cases of disease are rarer than the normal population [22].

3.3 Intrusion detection in networks

Network-based computer systems are increasingly playing a vital role in modern societies. Attacks on computer systems and networks are growing. Different categories of network attacks exist; some are numerous, and others are rare. For example, the KDD-CUP'99 dataset contains four categories of network attacks: denial of service (DoS), monitoring (probe), root to local (R2L), and user to root. (U2R). The last two attacks are intrinsically rare [23].

4. Evaluation metrics

The classical performance measures used for evaluating the performances of classifiers when used with balanced datasets are not appropriate for imbalanced datasets. This is because they have a strong bias toward majority class and are sensitive to class skews [24–27]. For example, the accuracy measure is not appropriate for the problem of imbalanced datasets [28]. If we consider a dataset which contains only 1% of minority instances and 99% of majority instances, the accuracy is 99% if all majority instances are well classified. However, misclassified 1% minority instances may lead to an enormous cost, and 99% accuracy could be a disaster for a medical diagnosis. Consequently, other metrics are necessary for measuring the performances of classifiers.

Some measures are extracted directly from the confusion matrix. They measure the classification performance of the majority and minority classes independently. Some others are combined to measure the performance of a classifier. They are described below.

4.1 Precision

It is a measure of accuracy [29]. It represents the percentage of well-classified minority instances in relation to all instances whose predicted class is a minority. It is defined in Eq. (2):

$$Precision = \frac{TP}{TP + FP} \tag{2}$$

4.2 Recall

It is the percentage of minority instances which are well classified as belonging to the minority class. In literature, this metric has several names such as sensitivity, true positive rate (TPrate), or positive accuracy [30]. It is defined in Eq. (3):

$$Recall = \frac{TP}{TP + FN} \tag{3}$$

4.3 Specificity

It is the percentage of majority instances which are well classified as belonging to the majority class. This measure is also known as true negative rate (TNrate) or negative accuracy. It is defined in Eq. (4):

$$Specificity = \frac{TN}{TN + FP} \tag{4}$$

4.4 False-positive rate (FPrate)

It is the percentage of majority instances misclassified as belonging to the minority class. It is defined in Eq. (5):

$$FPrate = \frac{FP}{FP + TN} \tag{5}$$

4.5 False-negative rate (FNrate)

It is the percentage of minority instances misclassified as belonging to the majority class. It is defined in Eq. (6):

$$FNrate = \frac{FN}{FN + TP} \tag{6}$$

4.6 G-mean

It indicates the balance between classification performances on the majority and minority classes [30]. A poor performance in the prediction of the positive instances will lead to a low G-mean value even if the negative instances are correctly classified by the model [31]. It has been used by several researchers for evaluating classifiers on imbalanced datasets [31–33]. G-Mean takes recall and specificity into account simultaneously. It is defined in Eq. (7). This metric will be used to test our approach:

$$\mathrm{G-Mean} = \sqrt{Recall * Specificity} \tag{7}$$

4.7 F-measure

It is defined as the harmonic mean of precision and recall [34]. Its value increases proportionally with the increase of precision and recall; a high value of F-measure indicates that the model performs better on the minority class. This metric is defined in Eq. (8):

$$\mathrm{F-Measure} = \frac{2 * Recall * Precision}{Recall + Precision} \tag{8}$$

4.8 Receiver operating characteristic curve (ROC)

The ROC curve [34, 35] is a technique for visualization, organization, and selection of classifiers based on their performances. It has long been used in signal detection to represent the trade-off between the success rate and false alarm rate of classifiers. It is a two-dimensional graph where TPrate is plotted on the y-axis and FPrate is plotted on the x-axis.

For a discrete classifier, the pair (FPrate, TPrate) is produced that corresponds to one point in the ROC space. However, a probabilistic classifier produces a continuous numerical value. Therefore, a threshold may be used to produce a series of points in the ROC space to produce a curve instead of one point.

4.9 Area under the ROC curve (AUC)

From the ROC curve, we define another measure called area under the curve (AUC) [35, 36] defined in Eq. (9) to compare the performance of two classifiers. If the area associated with classifier C1 is greater than that associated with classifier C2, then the performances of C1 are better than C2:

$$AUC = \frac{TPrate + TNrate}{2} = \frac{1 + TPrate - FPrate}{2} \tag{9}$$

5. Approaches and techniques

The several approaches have been proposed to handle the classification problem in imbalanced datasets. These approaches are divided into three levels [20]: data level, algorithm level, and hybrid level.

5.1 Data level methods

It consists of resampling the data in order to decrease the effect caused by the imbalance [3]. They are classified into three groups [3]: oversampling, undersampling, and hybrid methods.

5.1.1 Oversampling methods

Oversampling is used to increase the size of an imbalanced dataset by duplicating some minority instances. This duplication can be done by the following methods.

5.1.1.1 Random oversampling

It duplicates some minority instances chosen randomly [3]. Therefore, the multiple copies of minority instances increase the overlapping between these instances [37]. In particular, the overlapping appears when the produced classifier contains more specific rules for multiple copies of the same instance. As a result, the accuracy of learning is high in this scenario, and the performance of the classifier for the test is generally low [38].

5.1.1.2 Synthetic minority oversampling technique (SMOTE)

SMOTE [39] is a synthetic method with data generation. It has achieved several successes in various fields [3]. It creates a synthetic example x_{new} for each minority instance x_i as follows. It determines the K-nearest neighbors (which are minority instances whose Euclidean distance between them and x_i is the smallest) of x_i. Then, it selects randomly one of K-nearest neighbors y_i. Finally, it applies Eq. (10), where δ is a random number $\in [0, 1]$. Therefore, we understand that x_{new} is a point of the segment joining x_i and y_i:

$$x_{new} = x_i + \left(y_i - x_i\right) * \delta \tag{10}$$

SMOTE will not ignore the minority instances because it generalizes decision regions for them. But SMOTE has two problems [40]: overgeneralization and variance. The first problem is due to the blind generalization of the minority area without taking into account the majority class, which increases the number of overlapping between classes. The second problem concerns the number of generated synthetic instances which is set in advance without taking into account the IR.

5.1.1.3 MSMOTE

SMOTE does not consider the distribution of minority instances and those that are noisy in a dataset. For this reason, MSMOTE [41] divides the minority instances into three groups: security, border, and latent noises.

An instance is secretary, if the number of its K-nearest neighbors belonging to the minority class is greater than those belonging to the majority class.

An instance is border, if the number of its K-nearest neighbors belonging to the minority class is lower than those belonging to the majority class.

An instance is latent noise, if all its K-nearest neighbors have the majority class.

MSMOTE generates synthetic instances for all security instances in the same way as SMOTE. However, for each border instance, it selects the most nearest neighbor to generate a synthetic example. But, it does not generate synthetic instances for noisy instances, because they decrease the classifier's performances.

5.1.1.4 Borderline-SMOTE

Border instances and those nearby are more likely to be misclassified than those that are far from the border, and they are the most important for classification. Based on this analysis, the border instances contribute little in the classification. Therefore, the Borderline-SMOTE [42] method has been proposed to apply oversampling to border minority instances instead of applying it to all minority instances. To do this, it constructs a set of border minority instances known as DANGER. Then, it applies SMOTE for each instance of the DANGER set.

5.1.1.5 Adaptive synthetic sampling approach (ADASYN)

ADASYN [43] uses a function called density as an automatic criterion to take a decision about the number of synthetic instances that may be generated of each minority instance.

5.1.2 Undersampling methods

It consists of reducing the data size by deleting some majority instances with the objective of equalizing the number of instances of each class [44]. There are several approaches of undersampling that differ in the way of selection of majority instances that will be deleted.

5.1.2.1 Random undersampling (RUS)

RUS [4, 44] removes some majority instances selected randomly. But it can potentially hinder learning [37, 38, 45]; the deleted majority instances can cause the classifier to ignore important concepts related to the majority class.

5.1.2.2 Informed undersampling

It is proposed to avoid the loss of information caused by RUS [46]. Among the algorithms of this kind of undersampling, we have the following.

5.1.2.2.1 EasyEnsemble

It aims to a better exploitation of majority instances ignored by RUS. At first, it divides the training dataset into minority set P and majority set N of sizes n and p, respectively [46]. Then, it builds T subsets N_1, N_2 ... , N_T of size p by applying random sampling with replacement on N. After that, it generates T classifiers H_1, H_2 ... , H_T. The classifier H_i is produced by applying AdaBoost on N_i and P, and it contains the concepts of all majority and minority instances. Finally, it constructs the final classifier H by combining the T generated classifiers.

5.1.2.2.2 BalanceCascade

The training dataset is composed of the sets P of minority instances of size p and N of majority instances of size n [46]. BalanceCascade constructs at each iteration the classifier H_i from all the set P and the subset E chosen randomly from N, with $|E| = p$. Then, it updates N by deleting all majority instances which are well classified by H_i. This algorithm explores the majority instances in a supervised way because the set of majority instances is updated after generation of each classifier.

5.1.2.2.3 Informed undersampling with KNN

This technique [44] is based on the distribution characteristics of data by applying KNN algorithm [47]. The following three methods of this technique have been proposed:

NearMiss-1 selects majority instances as follows:

- For each majority instance x_i.
- For each minority instance x_j: Computes the distance d_{ij} between x_i and x_j.
- Identify the three nearest neighbors x_k $(1 \le k \le 3)$ for x_i that represents minority instances.
- Compute the average distance d_i defined in Eq. (11):

$$d_i = \frac{1}{3}\sum_{k=1}^{3} d_{ik} \tag{11}$$

- Select majority instances x_i whose average distance to the three closest minority class instances is the smallest.

NearMiss-2 method has the same steps as the previous method. But, it selects the majority instances whose average distance to the three farthest minority class instances is the smallest.

NearMiss-3 selects a given number of the closest majority instances for each minority instance to guarantee that every minority instance is surrounded by some majority instances.

5.1.2.3 Undersampling with data cleaning techniques

The data cleaning techniques were applied to eliminate the overlapping between classes. In the following four subsections, we represent some methods.

5.1.2.3.1 Tomek links

Tomek links method [48] may be used as an undersampling method. It deletes the noisy majority instances and those that are close to the border. The obtained training dataset after removing the Tomek links is organized into set of clusters. This method may be used as a data cleaning technique to delete majority and minority instances.

5.1.2.3.2 Condensed nearest neighbor (CNN) rule

CNN [49] is an instance reduction algorithm proposed by Hart. It deletes the redundant majority instances. An instance is considered as redundant if it can be deduced from other instances. CNN uses the initial training dataset E to construct the consistent dataset E' that contains instances that correctly classify all instances of E using 1-NN algorithm. Its steps are:

1. Copy the first majority instance x and all the minority instances of the training dataset E into the sub dataset E'.

2. While there are misclassified instances in E, do:

    a. Classify the instance y (belonging to E) using E' and 1-NN.

    b. Add y to E', if it is misclassified.

CNN is sensitive to noise. However, noisy instances are more susceptible to be misclassified [50], and they will misclassify the instances of test dataset [50, 51].

5.1.2.3.3 Neighborhood cleaning rule (NCL)

NCL is an undersampling technique introduced by Laurikkala [52] to balance a dataset by applying data reduction. Its main advantage is that it takes into account the quality of the data with a focus on data cleaning more than reduction. It removes noisy majority instances using the edited nearest neighbor (ENN) algorithm, which is an instance reduction algorithm developed by Wilson [53]. It is used to delete all instances whose class differs at least twice from the class of its three nearest neighbors.

5.1.2.3.4 One-sided sampling (OSS)

OSS [17] is the result of using CNN followed by Tomek links. CNN is applied to remove redundant majority instances. However, Tomek links deletes the noisy majority instances and border minority instances.

5.1.2.4 Evolutionary undersampling (EUS)

EUS [20] results from the application of prototype selection [54] and genetic algorithm. It has eight models that depend on the objective that EUS aims to reach. For the first objective, there are two purposes. The first one is to balance a dataset without losing the accuracy, and then EUS is known as evolutionary balancing undersampling (EBUS). In the second one, EUS aims to obtain an optimal power of classification without taking in the consideration the balance of a dataset; it is called evolutionary undersampling guided by classification measures (EUSCM). For the second objective, there are two possibilities: majority selection (MS) instances only or global selection (GS) of both majority and minority instances.

5.1.3 Hybrid methods

These methods combine undersampling and oversampling. They aim to eliminate the overfitting [3] caused by oversampling methods. For examples, SMOTE+Tomek links [17] applies Tomek links after generation of synthetic minority instances by SMOTE, and SMOTE+ENN [17] uses ENN to delete minority and majority instances. For this, each misclassified instance of training dataset by its three nearest neighbors is deleted.

5.2 Algorithm level

Most approaches are based on either modifying the existing complete classification algorithms in order to adapt them to the imbalanced datasets or proposing specific ones.

5.2.1 Modification of the existing algorithms

5.2.1.1 Decision trees

A decision tree [55–58] is the most popular form of rule-based classifiers. It allows to model simply, graphically, and quickly a phenomenon more or less complex. Its readability, speed of execution, and the few necessary hypotheses a priori explain its current popularity. All the methods of constructing a decision tree have these operators: deciding if a node is terminal, selection of a test to associate to a node, and assignation a class to a leaf.

The existing methods of construction the decision trees differ by the choices made for different operators. CART [59] and C4.5 [60] are the most popular algorithms for decision trees.

In the construction phase of a tree, C4.5 selects the node attribute that maximizes the information gain [60], that is, a high value of confidence. However, this measure is not suitable for imbalanced datasets because the most confident rules do not imply that they are the most significant, and some of the most significant rules may not be the most confident (may not have high confidence). The same problem arises for CART, which uses the Gini function [60]. These algorithms focus on the antecedent to find the class. Also, they use sensitive measures to the class distribution. For these reasons, some approaches have been proposed which apply nonsensitive measures [61] or modify the construction phase.

For example, *class confidence proportion decision tree* (*CCPDT*) approach is a robust and insensitive approach. It generates rules that are statistically significant [62]. It focuses on each class to find the most significant antecedent. In this way, all instances are partitioned according to their classes. Therefore, the instances that belong to the different classes will not have an impact on the others. For this, the new class confidence (CC) measure has been proposed to find the most interesting antecedents of each class. It is defined in Eq. (12):

$$CC(X \rightarrow y) = \frac{Supp(X \cup y)}{Supp(y)} = \frac{TP}{TP + FN} \tag{12}$$

However, obtaining rules which have a high CC value is still insufficient to solve the problem. So, it is necessary to make sure that the classes implied by these rules not only have great confidence but are more interesting than their alternative classes. As a result, the new class confidence proportion (CCP) measure has been proposed. It is defined in Eq. (13):

$$CCP(X \rightarrow y) = \frac{CC(X \rightarrow y)}{CC(X \rightarrow y) + CC(X \rightarrow \overline{y})} \tag{13}$$

Therefore, the CCPDT approach modifies the C4.5 algorithm by replacing the entropy (the attribute partition criterion) by CCP.

5.2.1.2 Support vector machines (SVMs)

The Kernel-based learning methods are inspired by the statistical theory of learning and the dimensions of Vapnik-Chervonenkis (VC) [63], such as support vector machines (SVMs). These latter are supervised learning methods. They are used for classification in binary datasets in order to find a classifier that separates the data and maximizes the distance between these two classes. This classifier is linear, and it is called hyperplane. SVMs aim to find the most optimal hyperplane, which passes in the middle of the points of two classes and that maximizes the margin in order to minimize the classification error [64].

In imbalanced datasets, the ideal hyperplane is close to the majority instances, and the decision boundary is very close to the minority instances. In this case, the support vectors representing the minority instances are far from the ideal hyperplane. Thus, their contribution to the final hypothesis is little [32, 65, 66]. To solve this problem, several methods have been proposed that differ from the mechanism used. For example, in [67] the following three approaches were presented:

- Boundary movement (BM) that modifies the coefficient b in the kernel function.

- Biased penalties (BP) introduce different penalty factors for the minority and majority classes in objective function in the Lagrangian formulation. These factors reflect the importance of classes during the learning phase.

- Border class alignment (BCA) expands the border around the minority class much more than the border around the majority class.

5.2.2 Specific algorithms

The specific algorithms have been proposed to deal with classification problem in imbalanced datasets. Among them, we present RLSD and LUPC.

5.2.2.1 Rule learning for skewed datasets (RLSD)

RLSD [62] is an efficient algorithm for handling imbalanced datasets. Its discovery process leads from a specific search to a general search. First, it discovers rules for minority instances used in learning. Then, it compares them with majority instances. It has the following three research phases:

1. Discretization phase consists of dividing the values of a numerical attribute into a small number of intervals. Each interval is mapped by a discrete symbol.

2. Rule generation phase is a frequent data discovery process for minority instances. This algorithm summarizes this phase:

```
Input: the set of minority instances P and the maximum number of allowed rules
Output: the set of rules: Rules.
Begin
    1. Initially, the rule set is empty : Rules :=Ø;
    2. for each minority instance pєP do
      2.1. The rule R:=p;
      2.2. Consider R as an initial rule;
      2.3. If R does not belong to Rules then
          2.3.1. Merge R:
                For each rule RE є Rules:
                 If R and RE have common conditions then generate the new
                 rule NR with common conditions and apply the procedure
                 Add_Rules to add NR to Rules.
          2.3.2. Apply the procedure Add_Rules to add R to Rules.
    3. for each rule RE є Rules do if TPrate(RE)<Min_TPrate then delete RE
End.
```

The procedure Add_Rules adds the concerned rule R to the set of rules if it does not belong to this set. After that, if the number of rules exceeds M, then it deletes a rule selected randomly.

3. The evaluation and rule selection phase: in rule evaluation step, RLSD calculates the accuracy of each generated rule by the correspondence with each majority instance. A rule is deleted if its precision (defined in Eq. (3)) is less than the minimum precision. In rule selection step, RLSD selects the rule with the highest F-measure value (defined in Eq. (8)). Then, it deletes all minority instances covered by this rule. After that, F-measure is recalculated for the remaining rules using the rest of minority instances. This process is repeated until there are no more minority instances or there is no rule that covers the remaining minority instances.

5.2.2.2 Learning minority classes in unbalanced datasets (LUPC)

The main feature of LUPC [67] is the combination of the separate and conquer rule induction method [68] and the association rules [69].

It lets an imbalanced dataset of size D composed of the set of positive (minority) instances Pos and the set of negative (majority) instances Neg.

LUPC uses three measures of performances: accuracy [70] (acc), error rate (err) [64], and positive cover ratio (PCR). They are defined in Eqs. (14)–(16):

$$acc(R) = \frac{|Cov^{+}(R)|}{|Cov(R)|} \tag{14}$$

$$\text{err(R)} = 1 - \text{acc(R)} \tag{15}$$

$$PCR(R) = \frac{|Cov^{+}(R)|}{|D|} \tag{16}$$

where the coverage (Cov) [70] of the rule R is the percentage of instances that are covered by this rule. It is defined in Eq. (17):

$$Cov\ (R) = \frac{\text{number of covered instances}}{|D|} \tag{17}$$

$Cov^{+}(R)$ is the number of covered instances that have the same class as that of R.

A rule is *αβ-strong* if the conditions given in Eqs. (18) and (19) are checked, where parameters α and β are the thresholds, with $0 \le \alpha$ and $\beta \le 1$.

A rule is *non-αβ-forte* if the condition given in Eq. (20) is checked:

$$\text{acc(R)} \ge \alpha \tag{18}$$

$$\text{PCR(R)} \ge \beta \tag{19}$$

$$Cov^{-}(R) \ge \frac{1-\alpha}{\alpha} * Cov^{+}(R) \tag{20}$$

The steps of LUPC are:

Input: The sets Pos and Neg, the minimum threshold of accuracy : min_acc
and the minimum PCR: min_cov
Output: The set of rules: Rules_sets.
Begin
  Rules_sets:=Ø ;
  α, β :=**Initialize(Pos, min_acc, min_cov)** ;
  while (Pos ≠ Ø and (α, β)≠(min_acc, min_cov))
    Rule R := Best rule***(Pos, Nég, α ,β)*** ;
    If (R ≠Ø) then
      Pos := Pos – { instances covered by R }
      Rules_sets := Rules_sets U R ;
    else
      Reduce(α,β) ;
  Rules_sets:=Post traitement(Rules_sets) ; *// It is optional*
end.

The procedure "Initialize" depends on the user-specified bias on PCR or accuracy. It initializes α and β as min_acc and min_cov, respectively. Otherwise, α is initialized to 0.95 or min_acc if min_acc is greater than 0.95, and β is initialized to the maximum value of PCR of the attribute-value pairs available on the minority instances. To find the best rule, LUPC follows these steps:

1. Construct the set of attribute-value pairs: build the set E1 of all pairs, where each available pair (attribute, value) for positive instances is considered as a

part of condition whose class is C^+. Choose the set E2 (is a subset of E1) of the candidate pairs: each pair belonging to E1 is considered as a candidate pair if it covers more than $\alpha * \beta * |D|$ minority instances. Identify the αβ-strong pairs: each candidate pair will be checked on the Neg instances to see if it is αβ-strong. Order the αβ-strong pairs either by their precision or by their PCR. Choose η attribute-value pair candidates which are αβ-strong, and add them to the set of attribute-value pairs. In the case where the number of αβ-strong pairs < η, then add the pairs that are not αβ-strong and that have either a high accuracy or a high PCR.

2. Generate the set of candidate rules that contains γ rules belonging to the set of attribute-value pairs as follows:

   - Order all attribute-value pairs according to the accuracy and/or PCR.
   - If the number of αβ-strong pairs is greater than or equal to γ, then add γ pairs to the set of candidate rules otherwise:
     - First, put the αβ-strong pairs and the non-αβ-strong pairs in the set of candidate rules. Then, delete non-αβ-strong rules whose PCR is lower than β. After that, improve the set of candidate rules by iteratively executing the following procedure:
       1. Generate new rules by combining each non-αβ-strong rule of the set of candidate rules with the pairs which are in the set of attribute-value pairs.
       2. If the generated rules become αβ-strong, then they will be inserted in the first part of the set of candidate rules.
       3. The procedure stops if there is no change in the non-αβ-strong rules or the number of rules in the set of candidate rules is greater than γ.
       4. Reject the rules that satisfy the condition given in Eq. (20).

The values of α and β are gradually reduced by the rate Δa and Δc, respectively. The default quantities used are Δa = 2% and Δc = 1%.

5.3 Hybrid level

Some methods of the complete classification cannot deal with the classification in imbalanced datasets without being combined with other techniques. Among these methods, we present ensemble methods, the cost-sensitive learning, and some other approaches based on metheuristics.

5.3.1 Ensemble methods

Ensemble methods build a series of N classifiers and combine them to produce the final classifier C^* using voting strategies. They aim to obtain a high precision classifier. They are divided into two classes: boosting and bagging.

5.3.1.1 Boosting

Boosting algorithms [64] focus on difficult instances to classify without differentiating their classes. According to Prof. Zhou Zhi-Hua [71], the boosting algorithms are very efficient and able to deal with classification in imbalanced datasets, because the minority instances are likely to be misclassified, and therefore, they will have weights higher in the following iterations.

However, Mikel G. et al. [72] considered that the integration of data sampling methods can reduce the additional costs of automatically detecting the optimal distribution of representative classes and samples and also reduce the bias of a specific learning algorithm. Among these methods, we have SMOTEBoost, RUSBoost, and DataBoost-IM.

5.3.1.1.1 SMOTEBoost

It alters the distribution of the training dataset by adding minority instances generated by SMOTE [73] in order to provide it to the algorithm AdaBoost.M2 [74].

5.3.1.1.2 RUSBoost

It operates in a similar way to SMOTEBoost, but it applies random undersampling on the training dataset [75].

5.3.1.1.3 DataBoost-IM

It combines AdaBoost.M1 [76] with a data generation strategy [31]. It differs from the two previous algorithms, because it performs the balancing process for majority and minority instances after identifying the difficult instances. Its steps are as follows:

1. Produce the classifier C to detect the misclassified instances which are called seeds.

2. Order the seeds in ascending order of their weight.

3. Construct the sets MAJ and MIN, which contain M_j majority instances and M_m minority instances, respectively, that have the highest weights.

4. Generate N synthetic instances for each majority seed and M synthetic instances for each minority seed.

5. Update the weights taking into consideration the newly added synthetic instances.

5.3.1.2 Bagging

It constructs N classifiers on N distinct datasets [77]. Each dataset is known as bag; it is obtained by random sampling with replacement.

In imbalanced datasets, the number of majority instances in a bag is also high. The main factor to apply on bagging to adapt it to this kind of datasets is the way of collecting the instances. We distinguish three main algorithms in this family:

5.3.1.2.1 OverBagging

The distribution of instances may be taken into consideration in order to equalize the number of minority instances N_{min} and the number of majority instances N_{maj} [78]. Instead of constructing the bags randomly, we apply the oversampling according to the following two possibilities. In the first one, the minority instances are duplicated by oversampling, and majority instances are added directly, or they are selected by random sampling with replacement to increase the diversity. In the second one, the SMOTEBagging [78] is applied, where $A\%*N_{maj}$ minority instances are selected by random drawing with replacement and the remaining instances are generated by SMOTE. The factor A is called resampling rate. It is equal to 10% in the first iteration and 100% in the last (it is multiple of 10).

5.3.1.2.2 UnderBagging

The number of majority instances is reduced to the number of minority instances in each bag [79]. All minority instances may be in the same bag. But for increasing the diversity, they can be selected by random sampling with replacement.

5.3.1.2.3 UnderOverBagging

It follows the two previous methodologies, but it is identical to SMOTEBagging [79].

5.3.2 Cost-sensitive learning methods

Most classification algorithms ignore various misclassification errors and consider that all these errors have the same cost. In many real-world applications, this hypothesis is not true because the difference between different classification errors can be quite large.

Cost-sensitive learning methods [80, 81] have been given a lot of attention in recent years to address this problem. They have been divided into two categories: direct cost-sensitive learning [80] and cost-sensitive meta learning [80, 82]. They are also used for imbalanced datasets such as boosting and SVMs.

5.3.2.1 Cost-sensitive learning with boosting

In each iteration of boosting, the weights of misclassified instances increase by the same ratio whatever their classes. However, in imbalanced datasets, the number of misclassified minority instances is higher. Hence, it is necessary to distinguish between different sorts of instances in the weight attribution phase. Therefore, the higher weights may be attributed to the minority instances in order to be well classified. To achieve this goal, misclassification costs are introduced into the weight update equation. Among these algorithms [83, 84], we have AdaC1, AdaC2, and AdaC3. They differ in the way of introducing the misclassification costs into the weight update formula within the exponential part and into the equation of the calculation of classifier performance.

5.3.2.2 Cost-sensitive learning with support vector machines (SVMs)

SVMs [63] have been integrated with sampling methods to deal with the classification problem in imbalanced datasets. Among these methods, we have the following.

5.3.2.2.1 SMOTE with different costs (SDC)

SDC results from the application of SMOTE with different error costs (DEC) [32]. This method aims to shift the decision boundary far from minority instances and to increase their number. To achieve the first objective, Veropoulos et al. [85] proposed the use of different costs for the minority and majority classes. The minority instances are also duplicated by SMOTE to make them densely distributed in order to guarantee the most well-defined boundary.

5.3.2.2.2 Ensembles of over−/undersampling SVMs

These methods [86] balance the training dataset by preprocessing and providing it to SVM for building an optimal classifier. For instance, the ensemble of undersampling SVMs (EUS-SVM) applies SVM, N times on N different training datasets. It contains all minority instances and some majority instances selected by random sampling. The final classifier is built by the combination of N produced classifiers.

5.3.3 Approaches based on metheuristics

5.3.3.1 Undersampling by genetic algorithm (USGA)

This approach [87] applies an intelligent method to the extraction of the classification rules from imbalanced binary datasets based on three phases.

In phase 1, a learning algorithm is developed based on a genetic algorithm with the aim of extracting the first classifier noted C1, which covers only majority instances when available. Majority instances, which are well classified by the rules of C1, are removed. This approach balances the imbalanced dataset and prevents the loss of information contained in the deleted majority instances, which are replaced by the classification rules of C1. The number of deleted majority instances depends on the value of IR; the process is carried out until the IR is equal to 1. The genetic algorithm is used to find the "best" rule for the majority class from the imbalanced dataset. The quality of each rule is evaluated by satisfying specificity (defined in Eq. (4)). A rule is the best if it has a high number of well-classified majority instances.

In phase 2, the same procedure is applied to the obtained balanced dataset using a fivefold cross-validation to construct the classifier C2, which contains rules that represent both majority and minority instances. In this phase, the quality of each rule is evaluated by maximizing the accuracy (Eq. (14)).

In the third phase, they merge C1 and C2 to produce the classifier C3 at first, and then they process the obtained classifier C3 by eliminating the specific and contradictory rules.

5.3.3.2 ACOSampling

ACOSampling [13] is a method of undersampling based on ant colony optimization [88]. It handles imbalanced DNA microarray datasets. It consists of extracting the balanced dataset S' for the original dataset S as follows:

- For T times, it divides the dataset S into two datasets, training and validation.
- Each training dataset S_i is processed by applying modified ACO algorithm [88] to filter less informative majority instances and search the corresponding optimal training dataset S_i'.

- The statistical results from the T optimal training datasets are given in the form of a list of frequencies, where each frequency indicates the importance of the corresponding majority instance. The extracted instances are those with high frequency, which will be combined with all minority instances to construct the final balanced training dataset S'.
- It produces one classifier by support vector machines using S'.

6. Conclusion

In this chapter, we have presented the classification problem in imbalanced datasets, which are composed of two kinds of instances: majority instances and minority ones. We have also presented the different approaches and techniques used to handle this problem which are divided in three levels: data level, algorithm level, and hybrid level.

In future work, we are planning to present a state of the art about different approaches and techniques used to handle the classification problem in multi-class imbalanced datasets. Also, we will extend our proposed approach to this kind of datasets.

Author details

Aouatef Mahani* and Ahmed Riad Baba Ali
LRPE, FEI, University of Science and Technology of Algiers Houari Boumediene USTHB, Algiers, Algeria

*Address all correspondence to: mahani.aouatef@gmail.com

References

[1] Parpinelli RS, Lopes HS, Freitas AA. An ant colony based system for data mining: applications to medical data. In: Proceedings of the 3rd Annual Conference on Genetic and Evolutionary Computation (GECCO'01); 7-11 July 2001; San Francisco, California; 2001. pp. 791-797

[2] Lu H, Setiono R, Liu H. Effective data mining using neural network. IEEE Transactions on Knowledge and Data Engineering. 1996;**86**:957-961. DOI: 10.1109/69.553163

[3] Batista G, Prati RC, Monard MC. A study of the behaviour of several methods for balancing machine learning training data. ACM SIGKDD Explorations Newsletter. 2004;**61**:20-29. DOI: 10.1145/1007730.1007735

[4] Japkowicz N, Stephen S. The class imbalance problem: A systematic study. Intelligent Data Analysis. 2002;**65**:429-449. DOI: 10.3233/IDA-2002-6504

[5] Japkowicz N, Holte RC, Ling CX, Matwin S. Learning from Imbalanced Data Sets Workshop (ICML'2003). Washington, DC; 2003

[6] Weiss GM, Provost F. Learning when training data are costly: The effect of class distribution on tree induction. Artificial Intelligence Research archive. 2003;**191**:315-354. DOI: 10.1613/jair.1199

[7] Tang Y, Zhang Y, Chawla NV, Krasser S. SVMs modeling for highly imbalanced classification. IEEE Transactions on Systems, Man, and Cybernetics, Part B: Cybernetics. 2009;**391**:281-288. DOI: 10.1109/TSMCB.2008.2002909

[8] Alejo R, García V, Sotoca JM, Mollineda RA, Sánchez JS. Improving the performance of the RBF neural networks trained with imbalanced samples. Lecture Notes in Computer Science. 2007;**4507**:162-169. DOI: 10.1007/978-3-540-73007-1_20

[9] Fu X, Wang L, Chua KS, Chu F. Training rbf neural networks on unbalanced data. In: Proceedings of the 9th International Conference on Neural Information Processing (ICONIP'02), 18-22 November 2002; Singapore. Singapore: IEEE Xplore; 2003. pp. 1016-1020

[10] Murphey YL, Wang H, Ou G, Feldkamp LA. OAHO: An effective algorithm for multi-class learning from imbalanced data. In: IEEE International Joint Conference on Neural Networks (IJCNN 2007); August 12-17, 2007; Renaissance Orlando Resort. 2007. pp. 406-411

[11] Qiong G, Xian-Ming W, Zhao W, Bing N, Chun-Sheng X. An improved smote algorithm based on genetic algorithm for imbalanced data classification. Digital Information Management. 2016;**142**:92-103

[12] Yoon K, Kwek S. A data reduction approach for resolving the imbalanced data issue in functional genomics. Neural Computing and Applications. 2007;**16**:295-306. DOI: 10.1007/s00521-007-0089-7

[13] Yu H, Ni J, Zhao J. ACOSampling: An ant colony optimization-based undersampling method for classifying imbalanced DNA microarray data. Neurocomputing. 2013;**101**:309-318. DOI: 10.1016/j.neucom.2012.08.018

[14] Zhou ZH, Liu XY. Training cost-sensitive neural networks with methods addressing the class imbalance problem. IEEE Transactions on Knowledge and Data Engineering. 2006;**181**:63-77. DOI: 10.1109/TKDE.2006.17

[15] Ali K, Manganaris S, Srikant R. Partial classification using association

rules. In: Proceedings of the Third International Conference on Knowledge Discovery and Data Mining (KDD'97); 14-17 August 1997; Newport Beach, California. 1997. pp. 115-118

[16] Riddle P, Segal R, Etzioni O. Representation design and brute-force induction in a Boeing manufacturing domain. Applied Artificial Intelligence. 1994;**81**:125-147. DOI: 10.1080/08839519408945435

[17] Fernández A, García S, del Jesus MJ, Herrera F. A study of the behaviour of linguistic fuzzy rule based classification systems in the framework of imbalanced data-sets. Fuzzy Sets and Systems. 2007;**15918**:2387-2398. DOI: 10.1016/j.fss.2007.12.023

[18] Weiss GM. Mining with rarity: A unifying framework. ACM SIGKDD Explorations Newsletter. 2004;**61**:7-19. DOI: 10.1145/1007730.1007734

[19] Orriols-Puig A, Bernadó-Mansilla O, Goldberg DE, et al. Facetwise analysis of XCS for problems with class imbalances. IEEE Transaction on Evolutionary Computation. 2009;**135**:1093-1119. DOI: 10.1109/TEVC.2009.2019829

[20] García S, Herrera F. Evolutionary undersampling for classification with imbalance datasets: Proposals and taxonomy. Evolutionary Computation. 2009;**173**:275-306. DOI: 10.1162/evco.2009.17.3.275

[21] Ezawa K, Singh M, Norton SW. Learning goal oriented Bayesian networks for telecommunications risk management. In: Proceedings of the Thirteenth International Conference on Machine Learning (ICML'96); 3-6 July 1996; Bari, Italy; 1996. pp. 139-147

[22] Cohen G, Hilario M, Sax H, Hugonnet S, Geissbühler A. Learning from imbalanced data in surveillance of nosocomial infection. Artificial Intelligence in Medicine. 2006;**371**:7-18. DOI: 10.1016/j.artmed.2005.03.002

[23] Tavallaee M, Stakhanova N, Ghorbani A. Toward credible evaluation of anomaly-based intrusion-detection methods. IEEE Transactions on Systems, Man, and Cybernetics, Part C (Applications and Reviews). 2010;**405**:516-524. DOI: 10.1109/TSMCC.2010.2048428

[24] Daskalaki S, Kopanas I, Avouris N. Evaluation of classifiers for an uneven class distribution problem. Applied Artificial Intelligence. 2006;**205**:38-417. DOI: 10.1080/08839510500313653

[25] Huang J, Ling CX. Using AUC and accuracy in evaluating learning algorithms. IEEE Transactions on Knowledge and Data Engineering. 2005;**173**:299-310. DOI: 10.1109/TKDE.2005.50

[26] Landgrebe TCW, Paclick P, Duin RPW, Bradley AP. Precision-recall operating characteristic (P-ROC) curves in imprecise environments. In: Proceedings of the Eighteenth International Conference on Pattern Recognition (ICPR'06); 20-24 August 2006; Hong Kong, China: IEEE; 2006. pp. 123-127

[27] Provost F, Fawcett T. Analysis and visualization of classifier performance: Comparison under imprecise class and cost distributions. In: Proceedings of the Third International Conference on Knowledge Discovery and Data Mining (KDD'97); 14-17 August 1997; Newport Beach, California; 1997. pp. 43-48

[28] Joshi M. On evaluating performance of classifiers for rare classes. In: Proceedings of IEEE International Conference on Data Mining; 9-12 December 2002; Maebashi City, Japan; 2002. pp. 641-644

[29] Buckland M, Gey F. The relationship between recall and precision. American Society for Information Science. 1994;**451**:12-19. DOI: 10.1002/(SICI)1097-4571(199401)45:1<12::AID-ASI2>3.0.CO;2-L

[30] Kubat M, Matwin S. Addressing the curse of imbalanced training sets: One-sided selection. In: Proceedings of the Fourteenth International Conference on Machine Learning (ICML'97); 8-12 July 1997; Nashville, Tennessee; 1997. pp. 179-186

[31] Guo H, Viktor HL. Learning from imbalanced data sets with boosting and data generation: The Databoost-IM approach. ACM SIGKDD Explorations Newsletter. 2004;**61**:30-39. DOI: 10.1145/1007730.1007736

[32] Akbani R, Kwek S, Japkowicz N. Applying support vector machines to imbalanced data sets. Lecture Notes in Computer Science. 2004;**3201**:39-50. DOI: 10.1007/978-3-540-30115-8_7

[33] Wu G, Chang EY. KBA: Kernel boundary alignment considering imbalanced data distribution. IEEE Transactions on Knowledge and Data Engineering. 2005;**176**:786-795. DOI: 10.1109/TKDE.2005.95

[34] Fawcett T. An introduction to ROC analysis. Pattern Recognition Letters. 2006;**278**:861-874. DOI: 10.1016/j.patrec.2005.10.010

[35] Fawcett T. ROC Graphs: Notes and Practical Considerations for Data Mining Researchers. Technical Report HPL-2003-4. Palo Alto: HP Labs; 2003

[36] Egan JP. Signal Detection Theory and ROC Analysis. New York: Academic Press; 1975. 277 p

[37] Mease D, Wyner AJ, Buja A. Boosted classification trees and class probability/Quantile estimation. Journal of Machine Learning Research. 2007;**8**:409-439

[38] Holte RC, Acker LE, Porter BW. Concept learning and the problem of small disjuncts. In: Proceedings of the Eleventh International Joint Conference on Artificial Intelligence (IJCAI'89); 20-25 August 1989; Detroit, Michigan, USA; 1989. pp. 813-818

[39] Chawla NV, Bowyer KW, Hall LO, Kegelmeyer WP. SMOTE: Synthetic minority over-sampling technique. Artificial Intelligence Research. 2002;**16**: 321-357. DOI: 10.1613/jair.953

[40] Wang BX, Japkowicz N. Imbalanced data set learning with synthetic samples. In: Proceedings of IRIS Machine Learning Workshop; 09 June 2004; Ottawa, Canada. 2004

[41] Hu S, Liang Y, Ma L, He Y. MSMOTE: Improving classification performance when training data is imbalanced. In: Computer Science and Engineering, International Workshop (IWCSE'09); 28-30 October 2009; Qingdao, China; 2009. pp. 13-17

[42] Han H, Wang W, Mao B. Borderline-SMOTE: A new over-sampling method in imbalanced data sets learning. In: Proceedings of the International Conference on Intelligent Computing (ICIC'05); 3-6 August 2005; Nanchang, China; 2005. pp. 878-887

[43] He H, Bai Y, Garcia EA, Li S. ADASYN: Adaptive synthetic sampling approach for imbalanced learning. In: Proceedings of the International Joint Conference on Neural Networks; 1-8 June 2008; Hong Kong, China; 2008. pp. 1322-1328

[44] Zhang J, Mani I. KNN approach to unbalanced data distributions: A case study involving information extraction. In: Proceeding of International Conference on Machine Learning (ICML'03); 21-24 August 2003; Washington DC; 2003

[45] Drummond C, Holte R. C4.5, class imbalance, and cost sensitivity: Why under-sampling beats over-sampling. In: Proceeding of International Conference on Machine Learning (ICML'03); 21-24 August 2003; Washington DC; 2003. pp. 1-8

[46] Liu XY, Wu J, Zhou ZH. Exploratory under sampling for class imbalance learning. IEEE Transactions on Systems, Man, and Cybernetics, Part B (Cybernetics). 2006;**392**:539-550. DOI: 10.1109/TSMCB.2008.2007853

[47] Cover TM, Hart PE. Nearest neighbor pattern classification. IEEE Transaction on Information Theory. 1967;**131**:21-27. DOI: 10.1109/TIT.1967.1053964

[48] Tomek I. Two modifications of CNN. IEEE Transaction System, Man, Cybernetics. 1976;**611**:769-772. DOI: 10.1109/TSMC.1976.4309452

[49] Hart P. The condensed nearest neighbor rule. IEEE Transactions on Information Theory. 1968;**143**:515-516. DOI: 10.1109/TIT.1968.1054155

[50] Wilson DR, Martinez TR. Reduction techniques for instance-based learning algorithms. Machine Learning. 2000; **383**:257-286. DOI: 10.1023/A:1007626913721

[51] Aha DW, Kibler D, Albert MK. Instance-based learning algorithms. Machine Learning. 1991;**61**:37-66. DOI: 10.1007/BF00153759

[52] Laurikkala J. Improving identification of difficult small classes by balancing class distribution. Artificial Intelligence in Medicine. 2001;**2101**:63-66. DOI: 10.1007/3-540-48229-6_9

[53] Wilson DL. Asymptotic properties of nearest neighbor rules using edited data. IEEE Transactions on Systems Man and Cybernetics. 1972;**23**:408-421. DOI: 10.1109/TSMC.1972.4309137

[54] Ho SY, Liu CC, Liu S. Design of an optimal nearest neighbor classifier using an intelligent genetic algorithm. Pattern Recognition Letters. 2002;**2313**:1495-1503. DOI: 10.1016/S0167-8655(02)00109-5

[55] Witten IH, Frank E. Data Mining Practical Machine Learning Tools and Techniques. 2nd ed. San Fransisco, CA: Morgan Kaufmann Publishers; 2005. 560 p

[56] Mitchell T. Machine Learning. 1st ed. New York: McGraw-Hill Education; 1997. 432 p

[57] Adriaans P, Zantinge D. Data Mining. 1st ed. Harlow, England: Addison-Wesley Professional; 1996

[58] Tuffery S. Data Mining et Statistique Décisionnelle: l'intelligence dans les bases de données. 2nd ed. France: Editions Technip; 2005. 400 p

[59] Breiman L, Friedman JH, Olshen RA, Stone CJ. Classification and Regression Trees. 1st ed. London: Chapman and Hall/CRC; 1984. 368 p

[60] Quinlan JR. C4.5: Programs for Machine Learning. San Mateo, California: Morgan Kaufmann Publishers; 1993. 302 p

[61] Geisser S. The predictive sample reuse method with applications. American Statistical Association. 1975; **70350**:320-328. DOI: 10.2307/2285815

[62] Cieslak DA, Chawla NV. Learning decision trees for unbalanced data. Machine Learning and Knowledge Discovery in Databases. 2008;**5211**:241-256. DOI: 10.1007/11893028_93

[63] Michalewicz Z, Fogel DB. How to Solve it: Modern Heuristics. 2nd ed. Berlin Heidelberg: Springer-Verlag; 2004. 554 p. DOI: 10.1007/978-3-662-07807-5

[64] Schapire RE. The strength of weak learnability. Machine Learning. 1990;**52**: 197-227. DOI: 10.1007/BF00116037

[65] Raskutti B, Kowalczyk A. Extreme Re-balancing for SVMs: A case study. ACM SIGKDD Explorations Newsletter.

2004;**61**:60-69. DOI: 10.1145/1007730.1007739

[66] Wu G, Chang EY. Adaptive feature-space conformal transformation for imbalanced-data learning. In: Proceedings of the Twentieth on International Conference on Machine Learning (ICML'03); 21-24 August 2003; Washington, DC, USA. AAAI Press; 2003. pp. 816-823

[67] Ho TB, Nguyen D, Kawasaki S. Mining prediction rules from minority classes. In: Proceedings of the Fourth International Conference on Applications of Prolog (INAP2001); 20-22 October 2001; Tokyo, Japan; 2001. pp. 254-265

[68] Furnkranz J. Separate-and-conquer rule learning. Artificial Intelligence Review. 1999;**131**:3-54. DOI: 10.1023/A:1006524209794

[69] Agrawal A, Imielinski T, Swami A. Mining association rules between sets of items in large databases. In: Proceedings of the 1993 ACM SIGMOD International Conference on Management of Data (SIGMOD'93); 25-28 May 1993; Washington, D.C., USA; 1993. pp. 207-216

[70] Han J, Kamber M. Data Mining Concepts and Techniques. 2nd ed. San Francisco, CA, USA: Morgan Kaufmann; 2006. 800 p

[71] Zhou ZH. Ensemble Methods: Foundations and Algorithms. 1st ed. Florida: Chapman and Hall/CRC; 2012. 236 p. DOI: doi.org/10.1201/b12207

[72] Galar M, Fernández A, Barrenechea E, Bustince H, Herrera F. A review on ensembles for the class imbalance problem: Bagging, boosting and hybrid-based approaches. IEEE Transactions on Systems, Man, and Cybernetics, Part C: Applications and Reviews. 2012;**424**: 463-484. DOI: 10.1109/TSMCC.2011.2161285

[73] Chawla NV, Lazarevic A, Hall LO, Bowyer KW. SMOTEBoost: Improving prediction of the minority class in boosting. Knowledge Discovery in Databases. 2003;**2838**:107-119. DOI: 10.1007/978-3-540-39804-2_12

[74] Schapire RE, Singer Y. Improved boosting algorithms using confidence-rated predictions. Machine Learning. 1999;**373**:297-336. DOI: 10.1023/A:1007614523901

[75] Seiffert C, Khoshgoftaar TM, Van Hulse J, Napolitano A. RUSBoost: A hybrid approach to alleviating class imbalance. IEEE Transactions on Systems, Man, and Cybernetics - Part A: Systems and Humans. 2010;**401**:185-197. DOI: 10.1109/TSMCA.2009.2029559

[76] Freund Y, Schapire RE. A decision theoretic generalization of on-line learning and an application of boosting. Computer and System Sciences. 1997; **551**:119-139. DOI: 10.1006/jcss.1997.1504

[77] Breiman L. Bagging predictors. Machine Learning. 1996;**242**:123-140. DOI: 10.1023/A:1018054314350

[78] Wang S, Yao X. Diversity analysis on imbalanced data sets by using ensemble models. In: IEEE Symposium Series on Computational Intelligence and Data Mining (CIDM 2009); 30 March-2 April 2009; Nashville, TN, USA; 2009. pp. 324-331

[79] Barandela R, Valdovinos R, Sánchez R. New applications of ensembles of classifiers. Pattern Analysis and Applications. 2003;**63**:245-256. DOI: 10.1007/s10044-003-0192-z

[80] Turney PD. Cost-sensitive classification: Empirical evaluation of a hybrid genetic decision tree induction algorithm. Artificial Intelligence Research. 1995;**2**:369-409. DOI: 10.1613/jair.120

[81] Turney PD. Types of cost in inductive concept learning. In: Proceedings of the Workshop on Cost Sensitive Learning at the Seventeenth International Conference on Machine Learning (ICML'00); 29 June-02 July 2000; Stanford, California, USA; 2000. pp. 15-21

[82] Ling CX, Yang Q, Wang J, Zhang S. Decision trees with minimal costs. In: Proceedings of the Twenty-First International Conference on Machine Learning (ICML'04); 4-8 July 2004; Banff, Alberta, Canada; 2004. pp. 544-551

[83] Zadrozny B, Langford J, Abe N. Cost sensitive learning by cost-proportionate instance weighting. In: Proceedings of the Third International Conference on Data Mining (ICDM'03); 19-22 November, 2003; Melbourne, Florida, USA; 2003. pp. 155-164

[84] Sheng VS, Ling CX. Roulette sampling for cost-sensitive learning. In: Proceedings of the European Conference on Machine Learning (ECML-2007); 17-31 September 2007; Warsaw, Poland; 2007. pp. 724-731

[85] Veropoulos K, Campbell C, Cristianini N. Controlling the sensitivity of support vector machines. In: Proceedings of the International Joint Conference on Artificial Intelligence (IJCAI'99); 31 July 31-6 August 1999; Stockholm, Sweden; 1999. pp. 55-60

[86] Wang BX, Japkowicz N. Boosting support vector machines for imbalanced data sets. Foundations of Intelligent Systems. 2008;**4994**:38-47. DOI: 10.1007/978-3-540-68123-6_4

[87] Mahani A, Baba-Ali AR. A new rule-based knowledge extraction approach for imbalanced datasets. Knowledge and Information Systems. DOI: 10.1007/s10115-019-01330-9

[88] Colorni A, Dorigo M, Maniezzo V. Distributed optimization by ant colonies. In: Proceedings of the First European Conference on Artificial Life (ECAL'91); 11-13 December 1991; Paris. France; 1991. pp. 134-142

Chapter 5

Applications of Mining Arabic Text: A Review

Qasem Al-Radaideh

Abstract

Since the appearance of text mining, the Arabic language gained some interest in applying several text mining tasks over a text written in the Arabic language. There are several challenges faced by the researchers. These tasks include Arabic text summarization, which is one of the challenging open areas for research in natural language processing (NLP) and text mining fields, Arabic text categorization, and Arabic sentiment analysis. This chapter reviews some of the past and current researches and trends in these areas and some future challenges that need to be tackled. It also presents some case studies for two of the reviewed approaches.

Keywords: text mining, Arabic language, Arabic text categorization, Arabic sentiment analysis, Arabic text summarization

1. Introduction

The massive increase in the amount and availability of online documents made retrieving and searching for information a difficult job and a problem for web users. The need for efficient and powerful tools to automatically categorize or summarize text has emerged. To overcome this problem, several methods and techniques have been proposed. One of these solutions is using different text mining tasks. Basically, text mining can be defined as the process of extracting knowledge from the massive amount of textual data. For this purpose, researchers in the fields of information retrieval, natural language processing, and data mining have investigated several types of text mining tasks and methods. These tasks include text categorization [1, 2] and text summarization [3].

2. Features of the Arabic language

The Arabic language is a popular language and is a member of the Semitic language family. It is the most widely spoken language of almost 330 million people and as a native language in more than 25 countries in an area spread from the Arabian Gulf in the East to the Atlantic Ocean in the West. Arabic is a highly structured and derivational language, where morphology plays a very important role [4].

The Arabic language ranks fifth among the top 30 languages spoken worldwide.

The Arabic language has three diversities, which include classical Arabic, modern standard Arabic (MSA), and colloquial Arabic. The classical Arabic is usually used for religious and historical scripts. The MSA can be found in today's written Arabic text in most formal channels. The colloquial or dialectical Arabic is the spoken language in informal and social media channels. In addition, the dialects vary from one Arab country to another.

The Arabic language has its own script with 28 alphabet letters, and some letters have different shapes according to their position in the word. The Arabic text is written from right to left, and there is no capitalization for letters. The Arabic language includes three main parts of speech: noun, verb, and particle [5].

Arabic natural language and text mining applications must deal with several complex problems pertinent to the nature and structure of the Arabic language. For example, the tokenization process for the Arabic text is not a straightforward job because the language is morphologically rich and the words are compact, where a word can correspond to an entire phrase or sentence.

For these reasons, the Arabic language needs careful preprocessing since it has some features that are different from other languages. Besides, these challenges may affect the results of any text analysis process such as classification or sentiment analysis [6].

3. Arabic text categorization

In recent years, and with the rapid increase in the size of information on the Web, text categorization has attracted the attention of many researchers to use the text categorization (classification) as a way to simplify the access to useful information. Text categorization as one of the main text mining tasks can be defined as the process of assigning a predefined category (label) to an unlabeled document (text) based on its content [7]. Text categorization has been used for several applications such as improving the performance of information retrieval systems, as spam filtering, and in medical information systems [8].

In practice, the typical text categorization system includes four main phases, where each phase may further include several other steps. These phases include the text preprocessing phase, which has several steps that aim to prepare the text to be ready to be processed by the categorization model. Usually, this phase includes some other classic natural language processing techniques such as parsing, tokenization, stop word removal, and term weighting, stemming, and part of speech tagging.

Text tokenization is the process of breaking up a sequence of strings into pieces such as words, keywords, phrases, symbols, and other elements called tokens. The result of tokenization process is tokens or terms which make it easy to count the number of the same terms in each document or any other calculation process required for calculating the weight of each term in each document.

Stop word removal is the process of removing punctuation marks, formatting tags, digits, prepositions, pronouns, conjunction, and auxiliary verbs. Stop words can be defined as a set of words that contains high frequency in the document collection and will not help us in the categorization process. The removal of stop words has many advantages, such as reducing the size of the document collection and allowing for deciding the most frequent term in each document. The Arabic language includes several types of stop words. Besides, general words, numbers, and symbols and special characters are also considered as stop words in several applications. Arabic text may also contain words, letters, or sentences from another language such as English. Most applications removed them from the text before processing.

Stemming is defined as the process that returns the segment of the word left after removing some prefixes and suffixes from the word. This process is usually used to reduce the size of the term set of each document in the collection of training documents used in the categorization process [5].

Term selection is defined as the process of selecting the best terms to reduce the term space. In term selection, each term is assigned a weight value based on a weighting scheme in a given text, and then the text is represented as a vector of the weighted terms, and this weight value is used to select the most important terms.

Term weighting: in this process, the terms in the text are weighted using the vector space model. In the vector space model, each term in the text is assigned a weight value based on a method called term frequency-inverse document frequency (TF-IDF). This method is considered one of the most popular methods used to compute term weights [9]. In term frequency-inverse document frequency method, the TF refers to the number of occurrences of term i in a document j, whereas IDF can be computed as follows:

$$IDF_{i,j} = \log\left(\frac{N}{n_i}\right) \tag{1}$$

where N is the total number of documents in the collection; n_i is the number of documents in the collection that contain the term i.

After the term frequency (TF) and inverse document frequency (IDF) are computed for each term, the term weight ($W_{i,j}$) is obtained using the following equation:

$$W_{i,j} = TF_{i,j} \times IDF_{i,j} \tag{2}$$

3.1 Some related works for Arabic text categorization

In the last few years, the importance of text categorization for the Arabic language has attracted many researchers. In this section, we just briefly review some of them. For example, and starting from 2006, some works have used the K-nearest neighbor method for the categorization of Arabic text [10, 11]. Duwairi in [12] compared the three classifiers: Naïve Bayes (NB), K-nearest neighbor (K-NN), and distance-based algorithms for categorizing Arabic text; then Mesleh in [13] proposed an approach using support vector machine (SVM).

Duwairi et al. [14] applied the K-nearest neighbor algorithm for the categorization of Arabic text to compare among three stemming techniques: full stemming, light stemming, and word clusters.

Both Thabtah et al. [15] Noaman et al. [16] proposed approaches based on Naïve Bayesian method for Arabic text categorization. For this purpose, the authors used chi-square for feature selection and Naïve Bayesian for categorization. Gharib et al. [17] applied a support vector machine (SVM) to deal with the problem.

In 2011, several researches were published dealing with Arabic text categorization, and some of them used the same classical methods used before but for different datasets [18] and using different preprocessing methods such as stemming [19].

Some researchers provide a comparative study of some existing approaches. For example, the work presented in [20] compares the sequential minimal optimization (SMO), Naïve Bayesian, and decision tree (C4.5) algorithms to find the most applicable method for the classification of the stemmed and the non-stemmed Arabic text. Both [21] and [22] have proposed Hybrid approaches proposed that combined the K-nearest neighbor method and the binary particle swarm optimization. In the same year, an interesting approach based on association rule mining is proposed by Al-Radaideh et al. in [1].

In 2012, a neural network-based method was proposed by [23]. The authors used learning vector quantization (LVQ) algorithm and Kohonen self-organizing map (SOM) to find similarities in text for the purpose of categorization.

The researches continue proposing different methods to handle the categorization problem of the Arabic text. In 2015, a survey and a comparative study of several approaches were presented by [24]. Besides, Al-Radaideh and Al-Khateeb [8] applied the associative classifier to classifying Arabic articles from the medical domain. The experimental results reported by the authors showed that the associative classification approach outperformed the Ripper, SVM, C4.5 algorithms.

One of the most frequent challenges in automatic text categorization is the high dimensionality of terms that may affect the performance of the categorization model. A good solution to overcome this challenge is to use feature selection/reduction methods that allow selecting a subset of terms that best represent the whole text in the best way. Ghareb et al. [25] tackled the feature selection part of the text categorization problem where they presented a hybrid feature selection approach that combines several feature selection methods with an enhanced version of the genetic algorithms (GA). In the same direction, the authors of [2] proposed three enhanced filter feature selection methods (Category Relevant Feature Measure, Modified Category Discriminated Measure, and Odd Ratio2) for text classification.

3.2 Case study

One of the recent approaches that tackled the dimensionality reduction issue is proposed by Al-Radaideh and AlAbrat [26]. They proposed a method that used the term weighting and the reduct concept of the rough set theory to reduce the number of terms used to generate the classification rules that represent the rough set classifier.

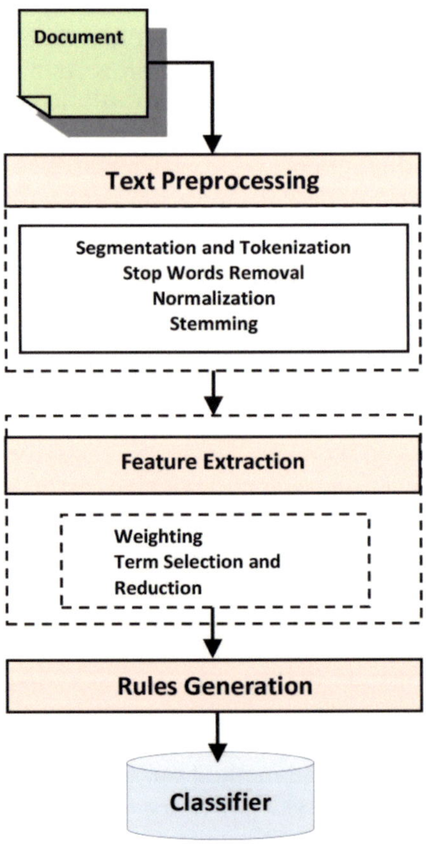

Figure 1.
The main phases of the classic text categorization process.

Category	F-measure for single reduct (%)	F-measure for multiple reducts (%)
Art	89	97
Economy	84	93
Health	94	98
Law	80	89
Literature	80	93
Politics	80	90
Religion	88	92
Sport	94	99
Technology	87	95

Table 1.
The F-measure for the two approaches.

The classification process starts with the classic preprocessing phase, which includes the tokenization step, stop word removal, term weighting, term stemming, and term selection. The main phases of the text categorization process are presented in **Figure 1**. The reduct concept was used to reduce the number of selected terms. The next phase is the rule generation phase. For this purpose, the authors enhanced the quick reduct method and proposed a multiple minimal reduct extraction method. The generated multiple reducts were used to generate the set of classification rules which represent the rough set classifier.

For evaluation purposes, an Arabic corpus of 2700 documents and 9 categories has been used. The documents in the corpus have been categorized manually by human experts into nine categories (art, economy, health, law, literature, politics, religion, sport, and technology). Documents in the corpus were evenly distributed on the 9 mentioned categories, with each having 300 documents.

Two experiments were reported, the first using single minimal reduct and the second using multiple reducts. The F-measure metric for the two methods is presented in **Table 1**. In addition, the reported results showed that the proposed approach had achieved an overall categorization accuracy of 94% when using multiple reducts, which outperformed the single-reduct method which achieved an accuracy of 86%.

4. Arabic sentiment analysis

The growth in the use of the social network's applications such as Facebook, Instagram, and Twitter has emerged into a huge volume of user reviews and opinions about the particular aspects of products or services where people like to share their opinions, feelings, experiences, thoughts, and preferences about these services, products, or events [27]. Reviewers express their feeling according to their understanding and observation.

In practice, this opinion can be used for identifying user interest and trends. One may have a positive opinion, while some others may have a negative opinion at the same time regarding a particular event or a service, and some express a neutral feeling.

Sentiment analysis is considered as one of the recent applications of text categorization that categorize the emotions expressed in a text [27]. To handle the huge amount of textual reviews and opinions, an intelligent automatic method is needed to analyze this huge content and classify it as positive, negative, or neutral opinion. In literature, this automatic process is called sentiment analysis (SA). Sentiment analysis is defined as the task that identifies the polarity of the review or the writer's attitude toward a particular topic, product, or a service is positive, negative, or neutral [28]. In recent years, this topic attracted many organizations where they use sentiment analysis to enhance the customer relationship management, improve the marketing process, and provide the client's feedback.

In practice, to analyze and classify sentiments of people is a difficult task to handle. There are several reasons for this issue. For example, the shared reviews and feelings are usually not in a structured format and written in a nonstandard language. Therefore, this analysis requires special techniques and semantic algorithms [29].

The science behind sentiment analysis is based on algorithms that use natural language processing concepts to categorize pieces of written text. The algorithms are usually designed to identify positive and negative words, such as "beautiful," "fantastic," "disappointing," and "terrible." There are mainly two levels of sentiment analysis, document level and sentence level.

Most sentiment analyzers work by one or a hybrid of the following four main approaches in scoring a sentence or document for the sentiment [30]:

1. Lexicon-based methods. These methods consider a lexicon dictionary for identifying the polarity of the text. Some predefined lists of positive and negative words were stored with their predefined scores (weights). The methods search for important keywords (usually verbs and adjectives) along with modifiers like negation words.

2. Rule-based methods. These methods look at the presence of some vocabulary words in sentences and use a set of predefined manually crafted rules to categorize these sentences by sentiment.

3. Machine learning methods. These methods treat the problem of sentiment analysis as a classification problem. The methods require some annotated datasets for training. The annotated datasets are lists of texts with sentiments manually recognized. If we take a massive amount of such texts and feed some machine learning algorithms (like Naïve Bayes, neural networks, SVM, deep learning) with them, it will learn how to identify or predict sentiment automatically. The machine learning methods usually involve some steps such as feature extraction from text and training and prediction processes.

4. Combined/hybrid methods. These methods use a lexicon dictionary along with pre-labeled dataset for developing a classification model using some machine learning approaches. Usually, by combining two approaches, the methods can improve the accuracy and precision of the sentiment analysis process.

4.1 Some related works to Arabic sentiment analysis

The majority of work on sentiment analysis has mainly targeted English text, whereas other languages such as the Arabic language have not received enough attention and focus. In recent years, several approaches were developed for Arabic sentiment analysis, where different works were started using different feature selection techniques and different dataset types from several different domains. Here we review some of these efforts.

For example, Abdul-Mageed et al. in [31] proposed a supervised machine learning system for sentiment analysis. They investigated different lexicon information to be either lemma or lexeme. The system used a sentence-level sentiment analysis. In the same year, Shoukry and Rafea [32] investigated using two classical machine learning methods: the Naïve Bayes and support vector machine classifiers.

In 2013, Al-Kabi et al. [33] investigated different classification methods for sentiment analysis. The methods include support vector machine and the Naïve Bayes. The TF-IDF term weighting method is used in the preprocessing step. Abdulla et al. in [34] investigated two main approaches, the corpus approach and the lexicon approach, by manually building the annotated corpus from the art domain.

Al-Subaihin and Al-Khalifa [35] provided a sentiment analysis system for Arabic text. Two algorithms have been designed to analyze and classify the text for their sentiment polarity. The system allows the users to annotate large Arabic text corpus using a game component. Then the system uses the sentiment analyzer component to classify the polarity of the text.

In 2014, some researchers investigated the effects of stemming, feature correlation, and n-gram models for Arabic sentiment analysis [6]. For the classification purpose, three classical classifiers were used, Naive Bayes, K-nearest neighbor, and SVM. In the same direction, Al-Radaideh and Twaiq [36] proposed to use the reduct concept of rough set theory as a term reduction method for the sentiment analysis of Arabic tweets. Both the number of the generated rules and then the number of reducts were used to measure the performance of the proposed method.

In 2015, a supervised ensemble-based classifier proposed by [37] combined different types of features such as opinion and discourse features. The classifier combines the three classifiers (MaxEnt, ANN, and SVM) with the majority voting for rules matching.

The standard Arabic corpus was always a problem for the researchers in the sentiment analysis domain. A prototype Arabic corpus was proposed, in 2016, by Al-Kabi et al. in [38]. The corpus could be used for sentiment analysis, Arabic reviews, and comments. The corpus consists of 250 topics related to 5 predefined domains.

Recently in 2017, Al-Radaideh and Al-Qudah [28] introduced how to use the concepts of rough set theory for term selection for sentiment analysis. The work considered an extension to work presented in [36]. The presented study used compared four reduct generation approaches and two rule generation methods. The results of the experiments showed that using rough set reduct techniques lead to different results, and some of them can perform better than non-rough set classifier. The conclusion of the work indicates that using the concepts of rough set theory for term selection/reduction can achieve good results.

Interesting work is presented by [39] where the authors used a standard corpus and made a review and performance comparison for some of the state-of-the-art approaches used in the multilingual sentiment analysis. One of the surprise results of the study is that the output accuracy of the reviewed approaches using the used standard corpus is far lower than the accuracies reported by the original research. This was justified due to the lack of details in the published works, which did not allow for exact reproduction of the reviewed methods. In some cases, the reported results by the original authors are not comparable with the results reported by [39] because they used different tools, experiment settings, and corpus.

After reviewing the results presented in some of the reviewed work, we could notice that no stable results were reported for the same methods used in the sentiment analysis process. This could be justified that the authors used different Arabic corpora with different dialectics and different stemming techniques and stop word lists.

5. Arabic text summarization

Text summarization is the process of producing a shorter version of a specific text. The main goal of automatic text summarization is condensing the source text into a shorter version (summary) preserving its information content and overall meaning. In practice, the process aims to extract the most important text segments (called summary) found in a single document or a set of related documents.

To generate a perfect summary and still convey the important information in the original text, a full understanding of the original full documents is critical. To understand the document is considered a hard task since it requires semantic and morphological analysis [3].

5.1 Categories of text summarization approaches

Text summarization methods can be categorized based on different criteria. These categories are summarized as follows:

1. The first category is based on the input of the summarization process. The category includes two types, single- and multi-document summarization. In the single-document summarization, only one document is used as an input to the text summarization process. The work of Suneetha and Fatima [40] is an example of this type. In the multi-document summarization, the input is more than one document (a cluster of related text documents), and the output is a single summary. The work of Kumar and Salim [41] is an example of this type.

2. The second category is based on the output of the summarization process, and here also we have two approaches, extractive and abstractive. In the extractive approach, the extractive summary is created by copying significant units (usually sentences) of the original document. The extractive approach involves two main steps: in the first step, the sentences are assigned some scores on how important they are, while in the second step, the highest-ranking sentences are extracted and presented as a summary. An example of this type is the work of Gupta and Lehal [42].

   In the abstractive approach, an abstractive summary relies on the idea of understanding the original text and retelling it in fewer words; a new vocabulary is added, or even novel sentences are unseen in the original sources. Abstractive approaches require deep natural processing such as semantic representation, compression of sentences, and the reformulation and use of natural language generation techniques, which did not yet reach a mature stage today. An example of this type is the work of Lloret and Palomar [43].

3. The third category is based on the content of the text, and here, two subcategories were found, indicative summary and informative summary. The indicative summary does not substitute the source document; it presents an indication about the document purpose and approach to the user so that the reader can choose which of the original documents to read further. An example of this type is the work presented in [44]. The informative summary can substitute the original document(s) by including informative facts that were reported in the original document. The work of [45] is an example of this type.

4. The fourth category is based on the generality of the summary, where two subcategories were found, generic and query-based summary. In the generic approach, the summarization system creates a summary of a document or a set of documents taking into account all the information found in the document. The work introduced in [46] is an example of this type. For the query-based summary, which is also known as user-focused or topic-focused, user-oriented

summarization systems try to build a summary relevant only to the user query. The work introduced in [47] is an example of this type of summarization.

5. The last category is based on the purpose of the summarization process. The category concerns the possible uses of the summary and the potential readers of the summary. It is divided into two types: domain-independent and domain-dependent. Domain-independent approaches do not have any predefined limitations and can accept different types of text. This kind of summarization makes few assumptions about the audience or the goal for generating the summary; anyone may end up reading the summary. An example of this type is the work in [48]. For the domain-specific approach, the system only summarizes documents belonging to a specific domain such as the medical domain; these kinds of systems exert some limitations on the subject of documents. Such systems know everything about a special domain and use this information for summarization. An example of this type is the work of Al-Radaideh and Bataineh [3].

5.2 The importance and usage of text summarization

Automatic text summarization systems are very significant in various domains such as news, medical, oil and gas, legal, and political domains. A good example, where text summarization could be used in the political domain, is the work presented in [49]. The authors introduced an automatic summarization system that summarizes the proceedings of the European Parliament using word clouds.

In general, text summarization is a very important text analysis and processing task and can be used for several purposes and several tasks. These can be summarized in the following points:

1. Text summarization allows users to quickly find the specific information they are looking for within documents.

2. It is important for different applications in natural language processing (NLP) such as information retrieval, question answering, and text classification systems. These applications can save time and resources, having their actual input text in condensed forms [50].

3. It has become very important for assisting and interpreting text information in today's fast-growing information age. This is presented in many applications for summarization such as summarizing news to short message services (SMS) or wireless application protocol format (WAP) for mobile phone, email summary, government officials' messages, or information for people in business [42].

4. Search engines such as Google uses text summarization to present compressed descriptions of the search results, to help users to decide the relevant documents quickly [43].

5. Summarizing domain-specific text such as news articles and political documents can lead to several benefits, such as:

    - It allows people who are interested in the political domain such as politicians to use text summarization in making decisions and getting the needed information instead of reading the whole document.

- Journalists in newspapers and electronic media take a long time in the preparation of news reports and articles due to a large number of archival and online documents related to the subject.
- Web users who are browsing newspapers and electronic media are overwhelmed with political news every day. Text summarization can save their time by helping them to decide which news to read.
- Mobile companies can use text summarization to summarize urgent news and events to SMS or WAP format and text them to their clients to keep them up to date with the latest news.
- Workers in the news agencies can prepare news briefs with the help of the text summarization.
- Producers of television programs and shows can use text summarization in making their stories and reports.

5.3 Some work in text summarization

The work on text summarization started a long time ago and still attracts many researchers. Several text summarization methods for English and other languages have been proposed in the literature. Over time, attention has shifted from summarizing scientific articles to news articles, electronic mail messages, advertisements, and sentiment-based text.

Research in automatic text summarization has gained importance with the widespread use of the Internet and the rapid increase of online information. Over the years, numerous automatic text summarization approaches have been proposed for English and other languages. For example, Silla Jr. et al. [51] tackled the automatic text summarization task as a classification problem. They used machine learning-oriented classification methods to produce summaries for documents based on a set of attributes describing those documents. Another example is the work presented in [52], where the authors proposed a summarization approach based on GA for sentence extraction. The approach used GA to produce a good summary that is readable, cohesive, and similar to the topic of the document.

Recent research focus has shifted to domain-specific summarization techniques that utilize the available knowledge specific to the domain of text [53] where automatic text summarization techniques have been applied to various domains such as medical, political, news, and legal domains proving that adapting domain-relevant features could improve the summarization performance. For example, Chen et al. [54] have proposed an automated text summarization approach (known as AutoTextSumm) to summarize oil and gas drilling articles. The approach combines statistical features, domain keywords, synonym words, and sentence position to extract the most important points in the document.

Some researchers tackled the multilingual text summarization. For example, Litvak et al. [55] have proposed a single-document extractive summarization called MUSE (multilingual sentence extractor) to improve multilingual summarization. The authors applied genetic algorithms to define a perfect weighted linear combination among 31 text-scoring techniques.

Nandhini and Balasundaram [56] have proposed a genetic-based text summarization system to assist in reading difficulties. The system considers informative score, readability score, and sentence similarity scores to weight the sentence of the text.

5.4 Arabic summarization methods

The first Arabic text summarization system called (Lakhas) has been proposed and implemented by [57]. This system uses some sentence scoring features such as term frequency, sentence position, words in the title, and cue words. A weighted linear combination of these features is used to score sentences. The top-ranked sentences are extracted to form the final summary.

A query-based extractive summarization approach has been proposed by [58]. The approach makes use of the phrasal decomposition of the text where each sentence is ascribed a scoring function that is used to identify the most relevant sentences in the text.

Al-Radaideh and Afif [59] proposed an approach that depends mainly on nouns to indicate the importance of the sentence. The approach was originally proposed for the Korean language.

Sobh in [60] proposed an Arabic extractive text summarization system based on machine learning, which integrates Bayesian classifier and genetic programming (GP) classifier in an optimized way to extract the summary sentences.

Hammo et al. [61] proposed a hybrid approach for Arabic text summarization. The approach used heuristic methods to rank text segments by assigning weighted scores to text segments. They used the Arabic WordNet to identify the thematic structure of the input text to select the most relevant sentences.

Two extractive approaches were proposed by [62]. The first is a graph-based approach, where the text is represented as a graph. The sentences represent the nodes of the graph, and edges between nodes represent the similarity between sentences. The second is a hybrid approach that combines the first approach with a statistical-based model.

Imam et al. [63] proposed an ontology-based summarization system for Arabic document. The system is a query-based system where the generated summaries aim at the user's interest according to the user query.

Oufaida et al. [64] proposed a summarization technique based on a statistical approach for assigning scores, to get minimum redundant and maximum relevance terms. Al-khawaldeh and Samawi [65] proposed an Arabic text summarization approach based on lexical cohesion and text entailment (LCEAS).

Al-Taani and Al-Rousan [66] proposed a multi-document text summarization approach based on clustering techniques. The clustering depends on text semantic to extract relationship across the sentences in a group of related documents.

5.5 Case study

One of the recent proposed approaches is presented by Al-Radaideh and Bataineh [3]. The authors proposed a hybrid, single-document text summarization approach that incorporates domain knowledge, statistical features, and genetic algorithms to extract important sentences from the Arabic text. The experimented domain was the political domain. The approach is tested using two corpora, the KALIMAT corpus and Essex Arabic Summaries Corpus (EASC).

KALIMAT corpus is a multipurpose Arabic corpus which contains 20,291 Arabic articles that were collected from the *Al Watan* newspaper. The articles in this corpus are divided into different topics, including political newswires. For each article there are two different summaries summarized by two human experts. The Essex Arabic Summaries Corpus (EASC) contains 153 Arabic articles collected from Wikipedia and *Al Ra'i* and *Al Watan* newspapers. The dataset contains 10 main topics including politics. For each document, five model extractive summaries are available.

The approach followed the typical main phases and steps of summarization systems. These phases are depicted in **Figure 2**. It can be noticed that some of the steps such as tokenization and stemming are general steps and are used across most text mining tasks such as text categorization and sentiment analysis.

In the approach of [3], the summarization process starts by passing the text into some preprocessing steps, which include text tokenization, stemming, and part of speech tagging. To evaluate and give a score to a sentence (*si*) that may appear in the final summary, the approach used several metrics. These metrics include the domain knowledge score (*Dkw*), term frequency (*TF*), sentence length (*SLen*), sentence position (*SPos*), and sentence similarity to the document's title (*SSTitle*). This score is called the informative score of the sentence which is used to determine its importance.

$$Score\ (si) = Dkw\ (si) + TF\ (si) + SLen\ (si) + SPos\ (si) + SSTitle\ (si). \qquad (3)$$

Dkw (*si*) is the score assigned to a sentence based on the number of political keywords it contains by summing up the weights assigned to the keywords in the domain knowledge base.

In the summary generation phase, the final summary is generated using the informative cores, semantic similarity, and genetic algorithms. To extract a cohesive summary from the text, the approach used the cosine similar to discover how sentences are related to each other.

The Recall-Oriented Understudy for Gisting Evaluation (ROUGE) metric was used to compare the automatically generated summaries by the approach against the summaries generated by humans. The ROUGE metric relies on different

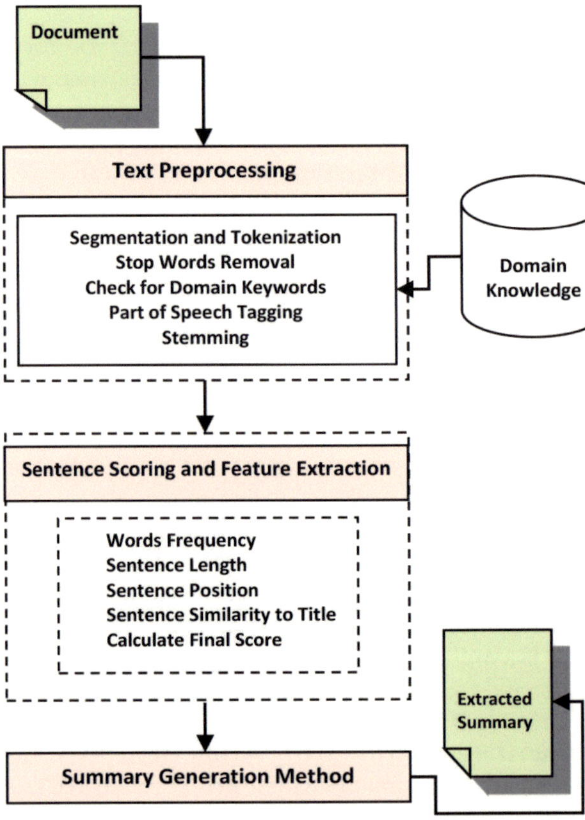

Figure 2.
The phases of the summarization process.

	Without domain knowledge		Using domain knowledge	
	CR = 25%	CR = 40%	CR = 25%	CR = 40%
Recall	36.6%	59.6%	31.9%	50.3%
Precision	62.6%	57.5%	60.7%	56.1%
F-measure	45.8%	60.5%	41.4%	52.8%

Table 2.
The ROUGE-1 evaluation results of the approach.

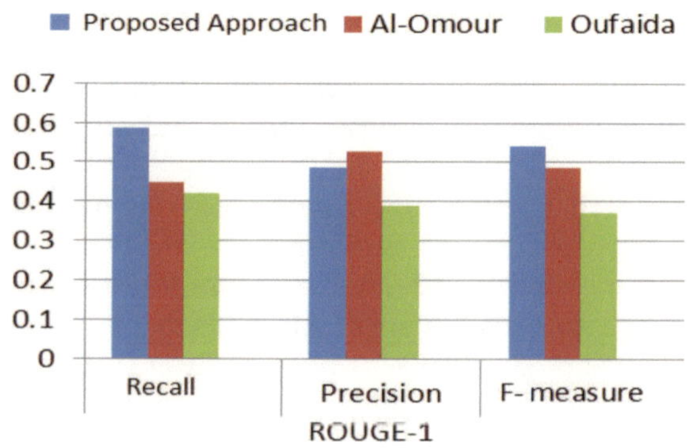

Figure 3.
Results of the three approaches.

types of n-gram co-occurrence. ROUGE-1 and ROUGE-2 compute the number of overlapping unigrams and bigrams, respectively, between the summary extracted by the summarization system and the golden summary created by humans. For each one of these metrics, the recall (R), precision (P), and F-measure (F) were calculated. Recall (R) is a measure of the coverage of the system (completeness). Precision (P) is a measure of correctness to find out how many sentences that the system returns are correct. F-measure (F) is calculated based on precision and recall to provide a single measurement for a summarizer. All the formulas are found in [3].

The approach is tested using two different compression ratios 25% and 40%. The approach is tested using the KALIMAT corpus with and without using the domain knowledge. The results are presented in **Table 2**. It can be noticed from the results presented in **Table 2** that the proposed approach using the domain knowledge achieved higher recall, precision, and F-measure values at both 25% and 40% compression ratios. The approach demonstrated promising results when summarizing Arabic political documents with an average F-measure of 60.5% at the compression ratio of 40%. This indicates the effect of using the domain knowledge corpus in the summarization process.

In the end, the approach is compared against two other Arabic text summarization approaches [62, 64] at a compression ratio of 40%. The results are presented in **Figure 3**.

6. Conclusion

This chapter discussed the three main text mining applications for Arabic language. These applications are text categorization, sentiment analysis, and text summarization. The chapter also reviewed some of the main related works to these applications and discusses some cases in detail. As a conclusion, although the process of Arabic text categorization was adopted by many researchers who have implemented several categorization algorithms, this field still needs more efforts to be enriched with new and improved algorithms. Till now, most proposed approaches used some well-known methods that could be used for different languages. In the future, new approaches that use the features of Arabic language need to be investigated. A real implementation of a complete Arabic text classification system is still a challenge.

As conclusion for sentiment analysis applications, we can say that some challenges still face the researches in this domain and need to be tackled. These challenges may apply to several languages, including Arabic. Sentiment analysis systems cannot understand perfectly the complexities of the human language. Recognizing context and tone is a difficult process for a machine. Beside, sentiment analysis is still rather incompetent in measuring things such as sarcasm, skepticism, hope, anxiety, or excitement.

Another challenge is that sentiment analysis needs to move beyond a one-dimensional positive to negative scale because there are other kinds of sentiment that cannot be placed on a simple scale. We need a multidimensional scale to truly understand and capture the broad range of emotions that humans express. The last challenge that should be noted is that sentiment analysis is highly domain centered. This means that a developed solution for one domain (e.g., mobile phones) will not directly work on other domains (e.g., hotels). The phrases and patterns used to express sentiment vary across domains and need to be adapted when switching between domains.

As for Arabic text summarization, it is still as one of the open, challenging areas for research in natural language processing (NLP) and text mining fields. Despite the existence of plenty of research work in the domain-based summarization in English and other languages, there is a lack of such work in Arabic due to the shortage of existing knowledge bases. This should motivate the researchers to develop automatic summarization approaches and applications to handle the increasing amount of electronic Arabic documents.

Author details

Qasem Al-Radaideh
Yarmouk University, Irbid, Jordan

*Address all correspondence to: qasemr@yu.edu.jo

References

[1] Al-Radaideh Q, Al-Shawakfa E, Ghareb A, Abu Salem H. An approach for Arabic text categorization using association rule mining. International Journal of Computer Processing of Languages. 2011;**23**(1):81-106

[2] Ghareb A, Bakar AA, Al-Radaideh Q, Hamdan A. Enhanced filter feature selection methods for Arabic text categorization. International Journal of Information Retrieval Research. 2018;**8**(2):1-24

[3] Al-Radaideh Q, Bataineh D. A hybrid approach for Arabic text summarization using domain knowledge and genetic algorithms. Cognitive Computation. 2018;**10**(4):651-669. DOI: 10.1007/s12559-018-9547-z

[4] Farghaly A, Shaalan K. Arabic natural language processing: Challenges and solutions. ACM Transactions on Asian and Low-Resource Language Information Processing. 2009;**8**:22. DOI: 10.1145/1644879.1644881

[5] Al-Kaabi M, Al-Radaideh Q, Akawi K. Benchmarking and assessing the performance of Arabic stemmers. Journal of Information Science (JIS). 2011;**37**(2):111-119

[6] Duwairi R, El-Orfali M. A study of the effects of preprocessing strategies on sentiment analysis for Arabic text. Journal of Information Science. 2014;**40**(4):501-513

[7] Lam W, Ruiz M, Srinivasan P. Automatic text categorization and its application to text retrieval. IEEE Transactions on Knowledge and Data Engineering. 1999;**11**(6):865-879

[8] Al-Radaideh Q, Al-Khateeb S. An associative rule-based classifier for Arabic medical text. International Journal of Knowledge Engineering and Data Mining. 2015;**3**(3-4):255-273

[9] Wang N, Wang P, Zhang B. An improved TF–IDF weights function based on information theory. In: Proceedings of the International Conference on Computer and Communication Technologies in Agriculture Engineering. 2010. pp. 439-441

[10] Al-Shalabi R, Kanaan G, Gharaibeh M. Arabic text categorization using KNN algorithm. In: Proceedings of the 4th International Multi-conference on Computer Science and Information Technology. Jordan: Amman; 2006

[11] Syiam MM, Fayed ZT, Habib MB. An intelligent system for Arabic text categorization. International Journal of Intelligent Computing and Information Sciences. 2006;**6**(1):1-19

[12] Duwairi R. Arabic text categorization. International Arab Journal of Information Technology. 2007;**4**(2):125-131

[13] Mesleh A. Chi-square feature extraction based SVMs Arabic language text categorization system. Journal of Computer Science. 2007;**3**(6):430-435

[14] Duwairi R, Al-Refai M, Khasawneh N. Feature reduction techniques for Arabic text categorization. Journal of the American Society for Information Science. 2009;**60**(11):2347-2352

[15] Thabtah F, Eljinini M, Zamzeer M, Hadi W. Naïve Bayesian based on chi-square to categorize Arabic data. In: Proceedings of the 11th International Business Information Management Association Conference (IBIMA) Conference on Innovation and Knowledge Management in Twin Track Economies, Cairo. 2009. pp. 930-935

[16] Noaman H, Elmougy S, Ghoneim A, Hamza T. Naïve Bayes classifier based Arabic document categorization. In: In: Proceedings of the 7th International Conference in Informatics and Systems (INFOS 2010); Cairo, Egypt. 2010

[17] Gharib TF, Habib MB, Fayed ZT. Arabic text classification using support vector machines. International Journal of Computers and Applications. 2009;**16**(4):1-8

[18] Al-Salemi B, Aziz M. Statistical Bayesian learning for automatic Arabic text categorization. Journal of Computer Science. 2011;**7**(1):39-45

[19] Wahbeh A, Al-Kabi M, Al-Radaideh Q, Al-Shawakfa E, Alsmadi I. The effect of stemming on Arabic text classification: An empirical study. International Journal of Information Retrieval Research. 2011;**1**(3):54-70

[20] Hussien MI, Olayah F, Al-dwan M, Shamsan A. Arabic text classification using SMO, Naive Bayesian, J48 algorithm. International Journal of Research and Reviews in Applied Sciences. 2011;**9**(2):306-316

[21] Chantar HK, Corne DW. Feature subset selection for Arabic document categorization using BPSO-KNN. In: Nature and Biologically Inspired Computing (NaBIC). 2011. pp. 545-551

[22] Chen Y, Zeng Z, Lu J. Neighborhood rough set reduction with fish swarm algorithm. Soft Computing. 2017;**21**(23):6907-6918

[23] Azara M, Fatayer T, El-Halees A. Arabic text classification using learning vector quantization. In: Proceedings of the 8th International Conference on Informatics and Systems (INFOS2012). 2012. pp. 39-43

[24] Hmeidi I, Al-Ayyoub M, Abdulla N, Almodawar A, Abooraig R, Mahyoub N. Automatic Arabic text categorization: A comprehensive comparative study. Journal of Information Science. 2015;**41**(1):114-124

[25] Ghareb A, Hamdan A, Bakar A. Hybrid feature selection based on enhanced genetic algorithm for text categorization. Expert Systems with Applications. 2016;**49**:31-47

[26] Al-Radaideh Q, AlAbrat M. An Arabic text categorization approach using term weighting and multiple reducts. Journal of Soft Computing. 2018;**2018**:1-15

[27] Rahmath H, Ahmad T. Sentiment analysis techniques–A comparative study. IJCEM International Journal of Computational Engineering & Management. 2014;**4**(17):25-29

[28] Al-Radaideh Q, Al-Qudah G. Application of rough set-based feature selection for Arabic sentiment analysis. Cognitive Computation. 2017;**9**(4):436-445

[29] Kumari U, Soni D, Sharma A. A cognitive study of sentiment analysis techniques and tools: A survey. International Journal of Computer Science and Technology. 2017;**8**(1):58-62

[30] Vohra M, Teraiya J. A comparative study of sentiment analysis techniques. Journal of Information, Knowledge and Research in Computer Engineering. 2013;**2**:313-317

[31] Abdul-Mageed M, Kübler S, Diab M. SAMAR: A system for subjectivity and sentiment analysis of Arabic social media. In: Proceedings of the 3rd Workshop on Computational Approaches to Subjectivity and Sentiment Analysis. 2012. pp. 19-28

[32] Shoukry A, Rafea A. Sentence-level Arabic sentiment analysis. In: Proceedings of International Conference on Collaboration Technologies and Systems (CTS); Denver. 2012. pp. 546-550

[33] Al-Kabi M, Abdulla N, Al-Ayyoub M. An analytical study of Arabic sentiments: Maktoob case study. In: Proceedings of 8th IEEE International Conference on Internet Technology and Secured Transactions (ICITST). 2013. pp. 89-94

[34] Abdulla NA, Ahmed NA, Shehab MA, Al-Ayyoub M. Arabic sentiment analysis: Lexicon-based and corpus-based. In: Proceedings of IEEE Jordan Conference on Applied Electrical Engineering and Computing Technologies (AEECT). 2013. pp. 1-6

[35] Al-Subaihin A, Al-Khalifa H. A system for sentiment analysis of colloquial Arabic using human computation. The Scientific World Journal. 2014;**2014**:8. Article ID: 631394. DOI: 10.1155/2014/631394

[36] Al-Radaideh Q, Twaiq L. Rough set theory approaches for Arabic sentiment classification. In: Proceedings of International Conference on Future of Things and Cloud, IEEE Computer Society. 2014

[37] Bayoudhi A, Hadrich L, Ghorbel B. Sentiment classification of Arabic documents: Experiments with multi-type features and ensemble algorithms. In: Proceedings of the 29th Pacific Asia Conference on Language, Information and Computation; Shanghai, China. 2015. pp. 196-205

[38] Al-Kabi M, Al-Ayyoub M, Alsmadi I, Wahsheh H. A prototype for a standard Arabic sentiment analysis corpus. The International Arab Journal of Information Technology. 2016;**13**(1A):163-170

[39] Dashtipour K, Poria S, Hussain A, Cambria E, Hawalah A, Gelbukh A, et al. Multilingual sentiment analysis: State of the art and independent comparison of techniques. Cognitive Computation. 2016;**8**:757-771

[40] Suneetha M, Fatima S. Corpus based automatic text summarization system with HMM tagger. International Journal of Soft Computing and Engineering (IJSCE). 2011;**1**(3):2231-2307

[41] Kumar Y, Salim N. Automatic multi document summarization approaches. Journal of Computer Science. 2011;**8**(1):133-140

[42] Gupta V, Lehal G. A survey of text summarization extractive techniques. Journal of Emerging Technologies in Web Intelligence. 2010;**2**(3):258-268

[43] Lloret E, Palomar M. Text summarization in progress: A literature review. Artificial Intelligence Review. 2010;**37**(1):1-41

[44] Saggion H, Lapalme G. Generating indicative-informative summaries with SumUM. Computational Linguistics. 2002;**28**(4):497-526

[45] Yih W, Goodman J, Vanderwende L, Suzuki H. Multi-document summarization by maximizing informative content-words. In: Proceedings of the 20th International Joint Conference on Artificial Intelligence (IJCAI); Hyderabad, India. 2007. pp. 1776-1782

[46] Gong Y, Liu X. Generic text summarization using relevance measure and latent semantic analysis. In: Proceedings of the 24th Annual International ACM SIGIR Conference on Research and Development in Information Retrieval, New Orleans, LA, USA. 2001. pp. 19-25

[47] El-Haj M, Kruschwitz U, Fox C. Experimenting with automatic text summarization for Arabic. In: Vetulani Z, editor. Human Language Technology. Challenges for Computer Science and Linguistics. LTC 2009. Lecture Notes in Computer Science. Vol. 6562. Berlin, Heidelberg: Springer; 2011

[48] Nomoto T, Matsumoto Y. The diversity-based approach to open-domain text summarization. Information Processing & Management. 2003;**39**(3):363-389

[49] De-Hollander G, Marx M. Summarization of meetings using word clouds. In: The Computer Science and Software Engineering (CSSE) CSI International Symposium; Tehran. 2011. pp. 54-61

[50] Pal A, Maiti P, Saha D. An approach to automatic text summarization using simplified Lesk algorithm and Wordnet. International Journal of Control Theory & Computer Modeling (IJCTCM). 2013;**3**(4):15-23

[51] Silla CN, Pappa GL, Freitas AA, Kaestner CAA. Automatic text summarization with genetic algorithm-based attribute selection. In: Lemaître C, Reyes CA, González JA, editors. Advances in Artificial Intelligence—IBERAMIA. Lecture Notes in Computer Science, Vol. 3315. Berlin, Heidelberg: Springer; 2004

[52] Qazvinian V, Hassanabadi L, Halavati R. Summarising text with a genetic algorithm-based sentence extraction. International Journal of Knowledge Management Studies. 2008;**2**(4):426-444

[53] Yeh J, Ke H, Yang W, Meng I. Text summarization using a trainable summarizer and latent semantic analysis. Information Processing and Management. 2005;**41**(1):75-95

[54] Chen Y, Foong O, Yong S, Kurniawan I. Text summarization for oil and gas drilling topic. International Journal of Computer, Electrical, Automation, Control and Information Engineering. 2008;**2**(6):1799-1802

[55] Litvak M, Last M, Friedman M. A new approach to improving multilingual summarization using genetic algorithms. In: The 48th Annual Meeting of the Association for Computational Linguistics; Uppsala, Sweden. 2010. pp. 927-936

[56] Nandhini K, Balasundaram S. Use of genetic algorithms for cohesive summary extraction to assist reading difficulties. Applied Computational Intelligence and Soft Computing. 2013;**2013**:11. Article ID: 945623. DOI: 10.1155/2013/945623

[57] Douzidia F, Lapalme G. Lakhas, an Arabic summarization system. In: The Document Understanding Conference (DUC); Boston, USA. 2004. pp. 128-135

[58] Bawakid A, Oussalah M. A semantic summarization system: The University of Birmingham at TAC 2008. In: The First Text Analysis Conference (TAC); Maryland, USA. 2008. pp. 1-6

[59] Al-Radaideh Q, Afif M. Arabic text summarization using aggregate similarity. In: The International Arab Conference on Information Technology (ACIT'2009); Yemen. 2009. pp. 1-8

[60] Sobh I. An optimized dual classification system for Arabic extractive generic text summarization [M.Sc. thesis]. Giza, Egypt: Department of Computer Engineering, Cairo University; 2009

[61] Hammo B, Abu-Salem H, Evens M. A hybrid Arabic text summarization technique based on text structure and topic identification. International Journal of Computer Processing of Languages. 2011;**23**(01):39-65

[62] Al-Omour M. Extractive-based Arabic text summarization approach [M.Sc. thesis]. Irbid, Jordan: Department of Computer Science, Yarmouk University; 2012

[63] Imam I, Hamouda A, Khalek H. An ontology-based summarization system for Arabic documents (OSSAD).

International Journal of Computers and Applications. 2013;**74**(17):38-43

[64] Oufaida H, Nouali O, Blache P. Minimum redundancy and maximum relevance for single and multi-document Arabic text summarization. Journal of King Saud University Computer and Information Sciences. 2014;**26**(4):450-461

[65] Al-Khawaldeh F, Samawi V. Lexical cohesion and entailment based segmentation for Arabic text summarization (LCEAS). World of Computer Science and Information Technology Journal (WSCIT). 2015;**5**(03):51-60

[66] Al-Taani A, Al-Rousan S. Arabic multi-document text summarization. In: The 17th International Conference on Intelligent Text Processing and Computational Linguistics (CICLing 2016); Turkey. 2016

Section 3

Deep Learning and Its Applications

Chapter 6

Categorizing Patient Disease into ICD-10 with Deep Learning for Semantic Text Classification

Junmei Zhong and Xiu Yi

Abstract

How to leverage insights into big electronic health records (EHRs) becomes increasingly important for accomplishing precision medicine to improve the quality of human healthcare. When analyzing big Chinese EHRs, there are a lot of applications that we need to categorize patients' diseases according to the medical coding standard. In this paper, we develop natural language processing (NLP), deep learning, and machine learning algorithms to automatically categorize each patient's individual diseases into the ICD-10 standard. Experimental results show that the convolutional neural network (CNN) algorithm outperforms the recurrent neural network (RNN)-based long short-term memory (LSTM) and gated recurrent unit (GRU) algorithms, and it generates much better results than the support vector machine (SVM), one of the most popular conventional machine learning algorithms, demonstrating the great impact of deep learning on medical big data analysis.

Keywords: electronic health record, natural language processing, deep learning, convolutional neural networks, long short-term memory, gated recurrent unit, SVM, ICD-10

1. Introduction

It has been found out that the analysis of the big EHR can help accomplish precision medicine for patients to improve the quality of human healthcare. In EHR analysis, there are a lot of applications that need to categorize each patient's disease into the corresponding category with respect to a medical coding standard. The automatic categorization is very desirable for the massive EHR data sets. In this paper, we develop deep learning algorithms for semantic text classification of EHR. Our categorization system consists of four core components to accomplish the disease categorization. Firstly, by using domain knowledge of medical informatics and techniques of information fusion, we construct structured & meaningful patients' clinic profiles from the scattered and heterogeneous medical records such as inpatient and outpatient records, lab tests, treatment plans, and doctors' prescriptions for medications in the EHRs. This makes it possible for us to leverage the insights from the big data with artificial intelligence (AI) algorithms. Secondly, we extract each patient's historical disease descriptions in the clinic profiles and take each of them as a document for categorization. Thirdly, we use NLP algorithms for document tokenization and vector representation when necessary. The last component is to train the predictive model based on supervised classification to categorize each disease into one of the 26 categories according to the first-level disease categories of the 10th version of the International Classification of Diseases (ICD-10) and Related Health Problems to standardize medical records [1] through text data categorization using deep learning algorithms such as the CNN, LSTM, and GRU neural networks. Experiments of comprehensive studies show that the CNN algorithm outperforms the other deep learning algorithms, and it generates much better results than the traditional machine learning algorithms for the same data set according to the quantitative metric of F1-score.

Our contributions are demonstrated in the following two aspects:

- We construct the patients' individual clinic profiles from the scattered and heterogeneous clinic records and tables in real EHRs with our medical domain knowledge together with medical informatics for medical information processing. The constructed clinic profiles make it feasible for us to generate actionable intelligence from the unstructured EHR raw data sets using machine learning, NLP and artificial intelligence algorithms (AI).

- We design and train predictive models with NLP, embedded representations, and deep neural network algorithms to categorize patients' diseases into ICD-10 standard.

The remaining of this paper is organized as follows. The research methodology is discussed in detail in Section 2. The experimental results are presented in Section 3, and the conclusions of the paper are made in Section 4 together with some discussions and the direction of future work.

2. Research methodology

This research consists of 4 components for the categorization of EHRs: problem definition and data preparation and collection from EHR, text data extraction from the prepared and collected data, the tokenization of the Chinese documents using NLP, and supervised deep learning algorithms with embedded vector representations for tokens/words as inputs to the neural network architectures for the semantic categorization of each patient's disease symptom description into ICD-10 standard.

2.1 Problem definition and data collection

For the research of Chinese medical healthcare data analysis, we have obtained the Chinese EHRs from 10 Chinese hospitals in Shandong Province, China. These EHRs contain 179 scattered and heterogeneous clinic records and tables, for example, the patients' admission records, outpatient records, inpatient hospitalization records, all kinds of lab tests, prescriptions for medication information, surgery information, hospital and doctor information, patients' personal information and their family information. However, the data quality is not satisfactory because it is still at the early stage for the Chinese hospitals to create the EHRs for patients, and most of the doctors are more willing to write notes on their patients' record books rather than to type their notes in the computer systems. As a result, in the 179 clinic tables, only a portion of them contains useful information and there are too many non-filled columns in many tables. After the preprocessing process, only 85 tables are selected and even in these 85 tables, many records do not contain useful information and we need to do additional processing such as data governance [2] for some applications. We then construct individual clinic profiles for patients by using the method of database information fusion on the scattered and heterogeneous 85 tables according to the medical domain knowledge. The constructed clinic profiles contain each patient's individual symptom descriptions, diagnosis records, lab tests, doctor's treatment plans, and prescriptions for each office visit or each day's hospitalization, forming the clinic profiles. The clinic profile is very useful for all applications about the EHR analysis with machine learning. But for this categorization task, we only extract the disease symptom description of individual patients from the constructed clinic profiles, and then we use NLP, machine learning, and AI algorithms to analyze them. The number of valid records with non-empty symptom description in the EHR is significantly reduced after all preprocessing steps.

2.2 The tokenization of Chinese documents

When we use machine learning for document categorization, documents first need to be tokenized into individual words or tokens. For traditional machine learning algorithms, documents are represented as feature vectors with the bag of words (BOW) feature engineering method. For deep learning algorithms, each document is the input of the deep learning architecture with individual tokens represented as word embeddings. So, we first tokenize each Chinese document into a collection of terms. The tokenization of Chinese documents is very different from that of English, which can be accomplished through the delimiters between terms. In this work, we use the Han LP, an open source tokenization tool for tokenizing Chinese documents. After some preprocessing for the tokens, we then represent each word with the pre-trained embedded vector representation with the language model of distributed representation learning algorithm word2Vec [3].

2.3 Embedded vector representation for words

Both traditional one-hot vector representation and BOW-based TF-IDF or binary feature representation for words have a lot of limitations for document classification. They are constrained with problems of high dimensionality and sparsity for the feature vectors and they are not able to capture any semantic information from the words due to the mutual orthogonality of the individual words with such representations. The motivation of using the distributed embedded vector representation is to reduce the dimensionality of the vector while at the same time providing semantic information of words. The word2Vec algorithm [3] is a kind of distributed representation learning algorithm for the language model to capture each word's semantic information through the embedded dense vectors so that semantically similar words can be inferred from each other. There are two different models in the word2Vec algorithm and they are the continuous bag-of-words (CBOW) model and the skip-gram model as shown in **Figures 1** and **2**, respectively.

For a sentence, the CBOW model is used to predict the current word from its left side and right side context words, which are within a window centered at the current word. On the other hand, the skip-gram model is used to predict the surrounding context words in a sentence for the given current word with the context words within a window whose center is at the current word. The Word2vec model can be trained in two different ways: using the hierarchical softmax algorithm or using the negative sampling method. For the hierarchical softmax algorithm, we first get the vocabulary of the corpus and then we create a binary Huffman tree for all words according to their frequencies of occurrences, and all words are the leaf nodes of the Huffman tree. The main benefit of the Huffman tree is that it offers the convenient access of the frequent information. In the Huffman tree, the high-frequency words have short paths from the root of the tree to the individual leaf nodes, and the low-frequency

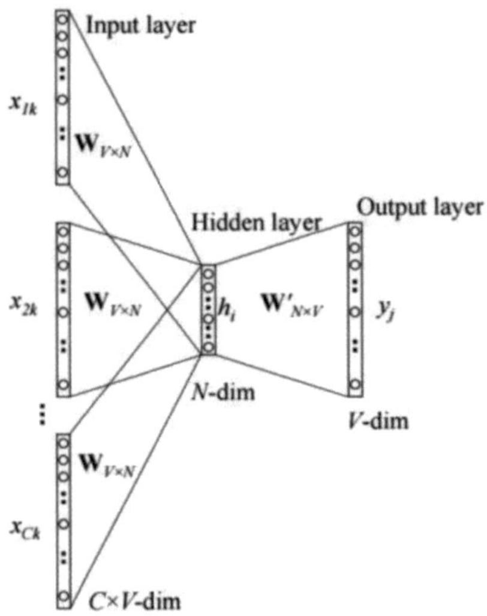

Figure 1.
The CBOW model in word2Vec.

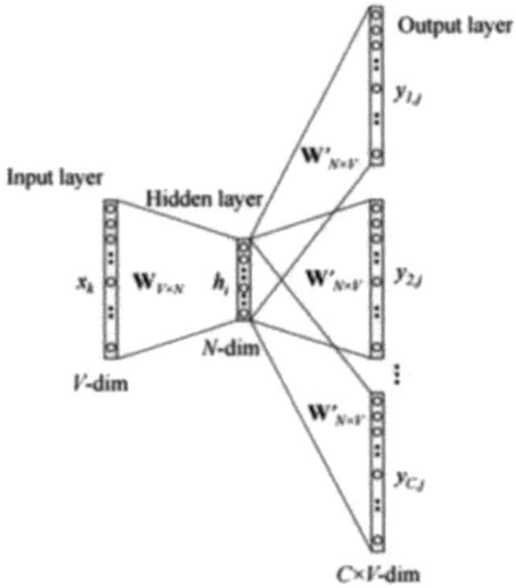

Figure 2.
The skip-gram model in word2Vec.

words have long paths, thus accomplishing optimal encoding performance from the viewpoint of bit-rate. In the original softmax algorithm, as shown in formula (1), when updating one word's vector during the training process, all words' vectors should be involved in the calculation and the computational complexity is accordingly as high as O(N). On the other hand, in the hierarchical softmax, when updating the embedded vector for a leaf node in the tree, only the leaf node and non-leaf-nodes on the path from the root of the tree to the leaf node need to be involved in the calculation. As a result, by using the Huffman tree based hierarchical softmax algorithm, we can significantly reduce the computational complexity from O(N) to O(logN).

$$softmax(x, w) = \frac{e^{x^T.w}}{\sum_j e^{x_j^T.w}} \tag{1}$$

As for the negative sampling, it does not need all words' vectors to be involved in updating a word's vector in the back-propagation process and it only needs to sample a few negative pairs like (central word, [negative context words]) for the central word of each positive pair (central word, positive context word) in the corpus. Furthermore, to improve the vector quality of low-frequency words, the high-frequency words are down-sampled and the low-frequency words are up-sampled by using a method called frequency lifting. This helps because otherwise the high-frequency words are more likely to be sampled than the low-frequency words for updating their vectors. With the embedded representation from Word2Vec, each word in the corpus can be represented by a unique low-dimensional dense vector. **Figures 3** and **4** demonstrate the clustering feature of this kind of word embeddings that if some words are semantically close to each other, their representations in the vector space are close to each other.

2.4 Deep learning for categorizing diseases into ICD-10

In this work, we build the predictive models based on deep learning techniques with the supervised learning methodology. To this end, we train four different deep learning models for performance comparison: the CNN with max pooling and k-max pooling, respectively, LSTM, and GRU.

2.4.1 The CNN model

The CNN is one of the popular deep neural network algorithms for both NLP and computer vision applications. The CNN architecture contains the layers of automatic feature extraction, feature selection, and the pattern classification. It has attracted great attention in text classification [4] and computer vision [5]. As demonstrated in **Figure 5**, different channels of the data can be used as inputs of the CNN architecture for feature extraction and feature selection via convolutional operations together with the pooling process and nonlinear activation. For computer vision applications, the R, G, B colors of an image are usually used as CNN's inputs. For text classification applications, a matrix of word embeddings stacked by the words' embedded vectors according to the order of the words in the sequence, is usually used as the input of the CNN architecture. The embeddings can be either from word2Vec, one-hot representation and/or other vector representations of words, forming different channels of the representation of text data. Within the CNN architecture, each channel of the texts is represented as a matrix, in which, the rows of the matrix represent the sequence of words according to their order, and each row is a word's vector representation with the number of columns being the

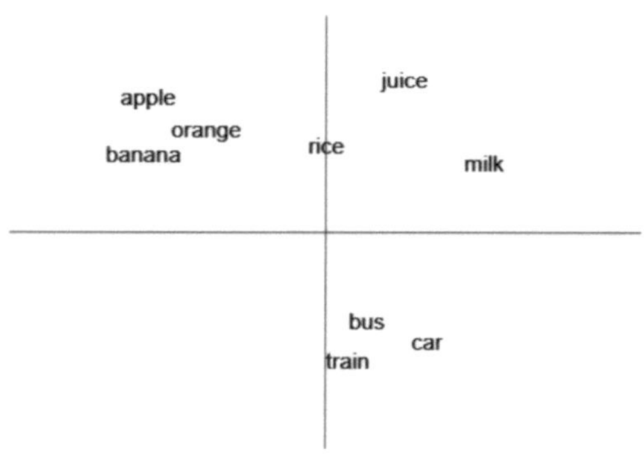

Figure 3.
The clustering characteristics of embedded vector representations of semantically similar words in the dense vector space.

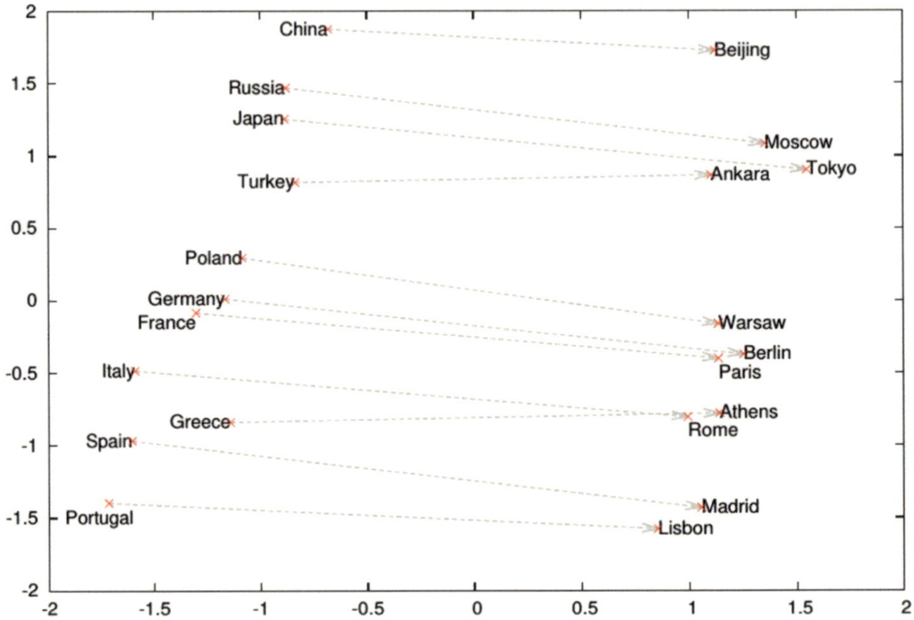

Figure 4.
The similar sub-structures of the embedded vectors in the dense vector space between semantically meaningful country-capital word pairs with the courtesy of Mikolov etc. [3].

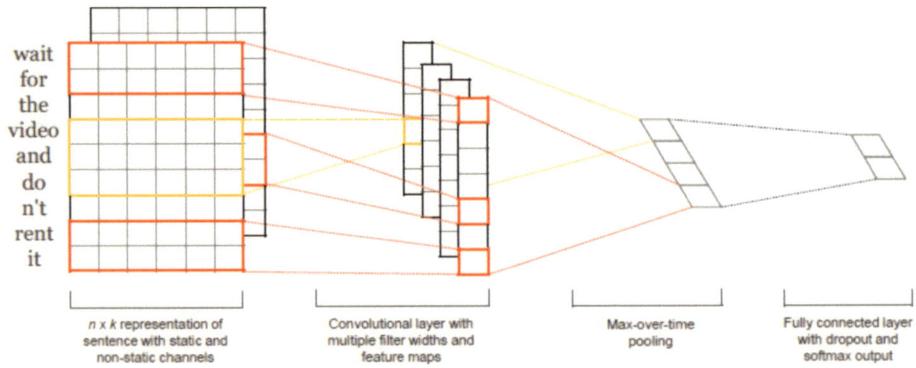

Figure 5.
Text classification using the CNN architecture with courtesy of Kim [4].

dimension size of the embedded vector space. For feature extraction, the matrix is convolved with some filters of different sizes, respectively. All filters have the same number of columns as the words' embeddings.

For text classification, the main idea of CNN using different sizes of filters is pretty much like that for computer vision. It is for the purpose of extracting the semantic features (patterns) of different N-Grams characterized by the filters. The different filter sizes correspond to different N-Grams. The words' vector representations can be either from the pre-trained word embeddings, for example, the word2Vec embeddings, or randomly initialized. For the randomly initialized words vectors, they are iteratively updated in the back-propagation process during the training stage until the training process is done, and this is another way of representation learning. For the clear illustration, we assume the filter w to have m rows, the maximum length of each sentence x is n, the dimensionality of word embeddings is d, then the sentence matrix and filter w can me mathematically formulated as $x \in R^{\wedge}(n \times d)$ and the filter $w \in R^{\wedge}(m \times d)$. CNN usually uses multiple such filters with different sizes m, for example, it uses 128 filters for each size m. During the convolution process, each of the filters gradually moves down by one word each time along the sequence of words. At each position, the filter covers a few (m) rows of the words' vector matrix, and the element-wise multiplication of the filter with the covered matrix is taken, and the multiplication results are summed up. This sum is then fed into the nonlinear rectified linear unit (Relu) activation function with an added biased term $b \in R$. The squeezed value is generated as a feature value, mathematically represented in the following formula:

$$f\_i = activation(w \cdot x[i{:}i + m - 1] + b) \quad (2)$$

where *activation* denotes the activation function such as Relu(), the dot operation between matrix w and x denotes the element-wise multiplication operation, and the subscript index i is the position index of the filter. After the convolution process is done for one filter for a sentence, a list of feature values is obtained like $feature\_map = [f\_1, f\_2, f\_3, \ldots, f\_(l - m + 1)]$. This is called a feature map corresponding to the filter. Finally, the pooling operation continues to take the most important feature value or a few most important feature values from the feature map as the output of the filter's convolution result with the sentence. When all filters are applied for convolution with the sentence's matrix, we obtain a feature vector for the input sentence. This feature extraction process and selection process makes the length of the final feature vector only dependent on the number of filters used for convolutions. The final step in the CNN architecture is a full connection layer including the dropout strategy and regularization, from the final feature vector to the output layer. The last layer of the CNN architecture is a full connection layer from the feature vector to the output. We finally obtain the classification result of a sample by applying the softmax function to the output for either binary or multi-class classification. For multi-class classification, the number of neurons of the output is equal to the number of classes to be predicted.

In addition to the commonly used max-pooling for CNN, another pooling method is the k-max pooling CNN [6], in which for each feature map obtained from the convolution of a filter w with the sequence matrix x, instead of selecting a single maximum value, it selects K consecutive maximum values. The benefit of using the k-max pooling strategy is to get some distribution information about the features in addition to the features themself in the feature map.

2.4.2 The RNN model

RNNs are very efficient for modeling the time series data for prediction and forecasting tasks. As shown in **Figure 6**, the chain-link RNN architecture consists of a sequence of neural network modules sharing the same parameters. Each module is used to analyze the information at a time in the sequence. At any time, the RNN system takes as inputs the information $x(t)$ at the specific time t, and the hidden state $h(t-1)$ of the previous module $c(t-1)$.

The typical NLP applications of using RNN are text classification, sentence generation, and language translation. With the sequence modeling, a sentence is taken to be an ordered time series with the starting time 0 at the first word and at time t, the corresponding word in the sentence has dependencies on its left-side words and its right-side words, which are considered as the contexts demonstrating the underlying syntactic and semantic information of the word in the sentence. For text classification with RNN, a sentence is usually encoded into a single fixed-length feature vector for classification, while for sentence generation and language

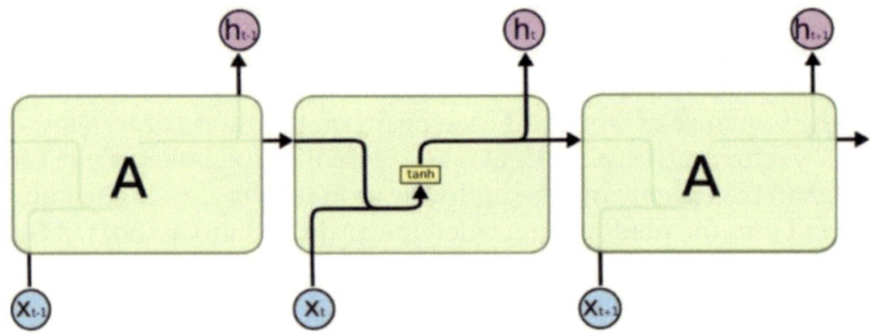

Figure 6.
The RNN structure with courtesy of Colah [7].

translation, the typical tasks are about predicting the next word given the context words seen so far together with what are generated or translated. The RNNs are great for prediction tasks that only need short contexts, but they no longer work for scenarios in which the long-term dependencies need to be remembered due to the intrinsic vanishing gradient problem of RNN during the back-propagation process. As a result, the long-term dependencies cannot be used for learning purposes and it is also very slow for the RNNs to be convergent. A significant tweaked version of RNNs, the long short-term memory (LSTM) networks [8], or the gated recurrent units (GRU) [9] is proposed to tackle the challenge of vanishing gradient.

2.4.3 The LSTM architecture

The LSTM networks are a special kind of RNNs, capable of learning the long-term dependencies in the input data. LSTMs are explicitly designed to avoid the vanishing gradient problem in learning the long-term dependencies in the data and they have achieved great success in sentiment analysis, text classification, and language translation. As shown in **Figure 7**, the LSTM architecture is very similar to that of the RNN. It is a chain of repeating modules, each of which is a modified version of that in RNNs.

By comparing a module at time t in **Figures 6** and 7, respectively, we can clearly observe that the LSTM architecture has some additional components in each module and these components are called gates with different functionalities to work together so that the long-term dependencies of words in the input sequence can be used for learning and the gradient vanishing problem can be solved. For NLP, both syntactic and semantic information can be encoded by LSTM networks for prediction and classification purposes. The first gate is the forget gate, and it is

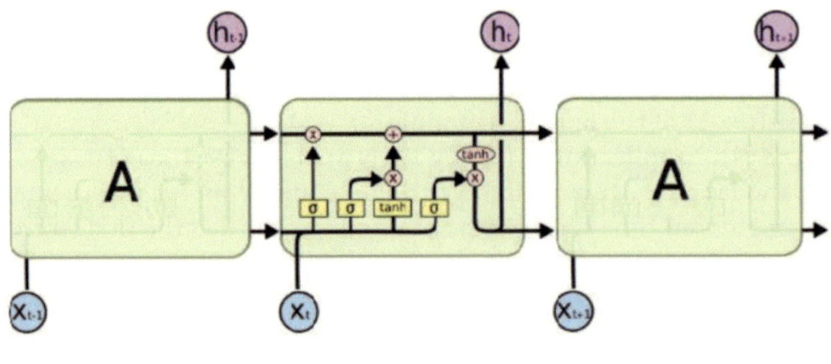

Figure 7.
The LSTM architecture with courtesy of Colah [7].

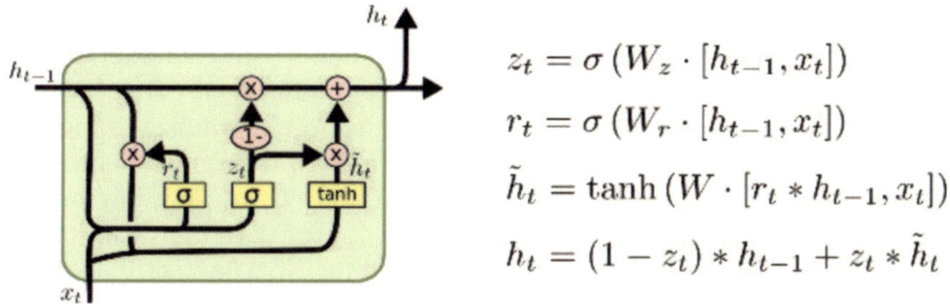

Figure 8.
The GRU structure with courtesy of Colah [7].

used to prevent some input information from entering into the module through a sigmoid layer to process $h(t-1)$ and $x(t)$. The output of the forget layer is a vector of values between 0 and 1, indicating how much the corresponding element in the old cell state $C(t-1)$ can be reserved to be the current cell state. If a value is 1, the corresponding information in the old state cell $C(t-1)$ is completely kept. On the other hand, if the value is 0, the corresponding element in the old state cell $C(t-1)$ is forbidden, and if the value is between 0 and 1, then only a portion of the corresponding element in the old state cell $C(t-1)$ is kept. The second gate is the input gate including a sigmoid layer and a tanh layer. The sigmoid layer is used to determine how much information in the inputs can be added to the current cell state $C(t)$, and the tanh layer creates a new vector of values to determine the polarity and proportion for the output of the sigmoid layer to be added to the current cell state $C(t)$. The elementwise product of the output of the sigmoid layer and the tanh layer is added to the current cell state $C(t)$. The third gate is the output gate which includes the filtered cell state $C(t)$ and a sigmoid layer. The output of the sigmoid layer with inputs of $x(t)$ and $h(t-1)$ is used to determine how much we are going to output from the inputs. The filtered cell state through a tanh activation function determines the polarity and proportion for each element in the current cell state $C(t)$ for updating the result of the sigmoid layer. Then the result of the filtered cell state is multiplied with the result of the sigmoid layer in an elementwise way to get the output of the current module as well as the hidden state $h(t)$ for the next module at time $t+1$. Also the current cell state $C(t)$ is transferred to the next module too.

2.4.4 The GRU architecture

The architecture of gated recurrent units (GRU) [9] is a simplified version of the LSTM architecture with only two gates, a reset gate r and an update gate z. As shown in **Figure 8**, the reset gate determines how to combine the input $x(t)$ with the previous memory $h(t-1)$, and the update gate defines how much of the previous memory can be kept. If the reset gate is set to be all 1's, and update gate is set to be all 0's, the GRU is reduced to be the original RNN model. Compared with LSTM, GRU has fewer parameters and thus may train the deep learning model in a faster way.

3. Experimental results and analysis

The ICD-10 coding system is essentially a tree-like hierarchical structure with 3 layers to encode patients' diseases. For a larger layer number in the tree, it categorizes diseases into finer disease categories. In the first layer, it only classifies disease symptoms into 26 coarse disease categories, but, in the second layer, it can classify disease symptoms into more than 500 disease categories, and in the third layer, it can accomplish the categorization for about 21,000 diseases. But for the time being, for supervised machine learning, we can only train a model to classify patient diseases into one of the 26 categories corresponding to the first layer of ICD-10 coding system. This is for the fact that as the number of classes increases for supervised machine learning, the required annotated training data increases significantly, however we do not have so many patient records now. Furthermore, our current EHR data is very unbalanced and most of the patients' diseases belong to popular diseases. For the 26 disease categories in the first-layer of ICD-10, our EHR system only has sufficient disease examples for 14 popular disease categories, but we do not have sufficient disease examples for the rest 12 unpopular disease categories. In this paper, we can only annotate disease examples for the 14 popular disease categories to train a 14-class classifier.

In this paper, we use the quantitative metrics to measure the performance of the models. They are the precision, recall, and F1-score, and they are calculated in the following way:

$$Precision = tp/(tp + fp) \quad (3)$$

$$Recall = tp/(tp + fn) \quad (4)$$

$$F1\_score = 2 \cdot (precision \cdot recall)/(precision + recall) \quad (5)$$

where *tp* denotes true positives, *fp* denotes false positives, and *fn* denotes false negatives.

For the fair performance comparison of deep learning algorithms with traditional machine learning algorithms, we compare the deep learning algorithms with what we have done with SVM [10]. For SVM, we use two kinds of vector representations for document representation: the BOW vectors and the embedded vectors. The BOW vectors can be further separated into TF-IDF weighted vectors and binary vectors. The embedded vectors can be separated into two forms: the averaged word embeddings of the pre-trained word embeddings from word2Vec to represent a document, and the doc2Vec vectors from the PV-DM model [11], respectively. As a result, we totally use 4 document representations as the inputs for SVM. For BOW method using both the TF-IDF weighted and binary vector representations, we use the keywords as the features instead of the individual tokenized words, and the dimensionality of the feature vectors is 53,039 after performing some preprocessing to filter out the infrequent keywords. When optimizing the word2Vec to get the pre-trained word embeddings, we have tried 4 models and they are the Skip-gram with the hierarchical softmax, Skip-gram with the negative sampling, CBOW with the hierarchical softmax, and CBOW with the negative sampling. We select the CBOW with the negative sampling to get the pre-trained word embeddings, and the feature vector of the SVM classifier is obtained as discussed above. The values for the optimized hyperparameters are: the number of epochs is set 10, minibatch is 32, dimension size of embeddings is 100, low frequency threshold for sampling is 1e−5, window size is 3, and 5 samples are used for negative sampling,

To train the multi-class SVM classifier, we use the grid search with the regularity ***L*1** and ***L*2**, respectively, to find the optimal hyperparameters. For both ***L*1** and ***L*2**, we use the hunger function to measure the empirical learning risk. The classification results of the 14-class SVM classifier are displayed in **Tables 1** and **2**, with respect to the regularity of ***L*1**, and ***L*2**, respectively. From the classification results of SVM, we can see that there is almost no difference between the regularity ***L*1**, and ***L*2**.

Input vector(s)	F1-score	Precision	Recall
Binary BOW	0.76	0.76	0.76
TF-IDF BOW	0.76	0.76	0.76
Avg. Word2Vec	0.70	0.71	0.71
Doc2Vec	0.64	0.65	0.65

Table 1.
The prediction results of 14-class SVM classifier with L1 regularity.

Input vector(s)	F1-score	Precision	Recall
Binary BOW	0.76	0.76	0.76
TF-IDF BOW	0.77	0.77	0.77
Avg. Word2Vec	0.70	0.71	0.71
Doc2Vec	0.64	0.65	0.65

Table 2.
The prediction results of 14-class SVM classifier with L2 regularity.

To train the multi-class CNN classifier, we use the pretrained word embeddings of word2Vec as the inputs of words' embeddings for the CNN architecture. For each filter size, we use 100 convolutional filters to extract features. The crawled Chinese medical documents are used as the corpus for training the word2Vec algorithm to get the pre-trained word embeddings. The optimized hyperparameter values of the CNN are: epoch size is 40, batch size is 20, dropout probability is 0.5, the $L2$-norm is used for regularization, and the dimension size of word embeddings is 200.

To train the multi-class LSTM and GRU classifiers, to make sure the deep learning models are trained in an optimal way, we have tried different optimization methods, such as the Adam, AdaDelta, and RMSprop. We find that the RMSprop optimizer works best for our data set. We set the hyperparameters to be: batch size 32, hidden size 64, epoch size 50, the dimension of word embeddings 300, dropout 0.5, and L2 regularization lambda 0.7. The LSTM and GRU algorithms are implemented in Python and Tensorflow. Since the training process for both LSTM and GRU is very time consuming for the use of 6000 training and validation examples for each disease category, for computational efficiency, we shorten each disease description to be 700 words. This assumes that customers' disease symptoms are mainly contained in the first 700 words of each note. Otherwise the memory requirement is very huge. This assumption is of course not 100% correct, but it greatly helps speed up the training process.

In this paper, the NLP, deep learning algorithms, and the algorithm of SVM are implemented in Python, Tensorflow, and the open source software of machine learning library Sklearn. The results of CNN, LSTM and GRU algorithms are displayed in **Table 3**, from which we can see that the CNN algorithm with max-pooling works best.

Input vector(s)	F1-score	Precision	Recall
Word2vec + CNN + k-max pool	0.74	0.73	0.73
Word2vec + CNN + Max pool	0.80	0.80	0.81
Word2vec + LSTM	0.74	0.75	0.74
Word2vec + GRU	0.74	0.75	0.74

Table 3.
The prediction results of 4 deep learning algorithms.

4. Conclusions and future work

In this paper, we develop NLP and deep learning algorithms to categorize patients' diseases according to the ICD-10 coding standard. Through comparative studies, we find out that the CNN model achieves better performance than the RNN-based LSTM and GRU models. The CNN model also outperforms the popular traditional machine learning model SVM for the same data set. In the future, we are going to investigate the transfer learning and deep learning algorithms with the attention mechanism for semantic text classification. At the same time, it is very necessary for hospitals and doctors to provide high-quality medical healthcare data and the high-quality EHR data is equally important as the medical services provided to patients.

Acknowledgements

The authors would like to thank the support from the Health Planning Commission of Shandong Province, China.

Author details

Junmei Zhong[1*] and Xiu Yi[2]

1 Marchex Inc, Seattle, WA, USA

2 Delta Technology Inc, Shandong, China

*Address all correspondence to: zhong.junmei@gmail.com

References

[1] ICD-10 Homepage. Available from: http://apps.who.int/classifications/icd10/browse/2016/en

[2] Junmei Z, Xiu Y. Artificial intelligence based data governance for Chinese electronic health record analysis. International Journal of Data Mining & Knowledge Management Process (IJDKP). 2018;**8**(3):29-41

[3] Tomas M, Kai C, Greg C, Jeffrey D. Efficient estimation of word representations in vector space. Advances in Neural Information Processing Systems. 2013:3111-3119

[4] Kim Y. Convolutional neural networks for sentence classification. In: Proceedings of the Conference on Empirical Methods in Natural Language Processing. 2014. pp. 1746-1751

[5] Krizhevsky A, Sutskever I, Hinton G. ImageNet classification with deep convolutional neural networks, NIPS'1. In: Proceedings of the 25th International Conference on Neural Information Processing Systems. Vol. 1. 2012. pp. 1097-1105

[6] Nal Kalchbrenner, Edward Grefenstette, Phil Blunsom. A convolutional neural network for modeling sentence. In: Proceedings of the 52nd Annual Meeting of Association for Computational Linguistics. Vol. 1. 2014. pp. 655-665

[7] Colah's Blog. Available from: http://colah.github.io/posts/2015-08-Understanding-LSTMs/

[8] Sepp H, Schmidhuber J. Long short-term memory. Neural Computation. 1997;**9**(8):1735-1780

[9] Chung J, Gulcehre C, Cho KH, Bengio Y. Empirical evaluation of gatedrecurrent neural networks on sequence modeling. arXiv:1412.3555v1

[10] Zhong J, Gao C, Yi X. Categorization of patient disease into ICD-10 with NLP and SVM for Chinese electronic health record analysis. In: Proceedings of 2018 International Conference on Artificial Intelligence and Pattern Recognition; Aug. 18-20, 2018; Beijing, China

[11] Le Q, Tomas M. Distributed representation of sentence and documents, proceedings of the international conference on. Machine Learning. 2014:1188-1196

Chapter 7

Deep Learning Approach to Key Frame Detection in Human Action Videos

Ujwalla Gawande, Kamal Hajari and Yogesh Golhar

Abstract

A key frame is a representative frame which includes the whole facts of the video collection. It is used for indexing, classification, evaluation, and retrieval of video. The existing algorithms generate relevant key frames, but additionally, they generate a few redundant key frames. A number of them are not capable of constituting the entire shot. In this chapter, an effective algorithm primarily based on the fusion of deep features and histogram has been proposed to overcome these issues. It extracts the maximum relevant key frames by way of eliminating the vagueness of the choice of key frames. It can be applied parallel and concurrently to the video sequence, which results in the reduction of computational and time complexity. The performance of this algorithm indicates its effectiveness in terms of relevant key frame extraction from videos.

Keywords: deep learning, neural network, histogram, video processing, computer vision

1. Introduction

In video analysis and processing, relevant and necessary information retrieval is a mandatory task, because if the video is large, then it is difficult to process the complete video in less time without losing its semantic details. Key frame extraction is a primary step of a computer vision algorithm. The key frame means the part of the video that can represent a visual summary and meaningful information about the video sequence. The key frame can be useful in many applications such as video scene analysis, browsing, searching, information retrieval, and indexing. Aigrain et al. in [1] describe the benefits of key frame extraction for information extraction in a video sequence. HongJiang et al. [2] significantly justify that for any video sequence the user can perform searching, indexing, and retrieval of information efficiently and faster using key frame extraction. Liu et al. [3] and Gargi et al. [4] proposed an object motion-based approach of key frame extraction. Basically, the video has a complex structure. It is a combination of the scene, shot and frames [5] as shown in **Figure 1**. In many computer vision applications such as content-based video retrieval (CBVR), video scene analysis and video sequence summarization is mandatory to analyze the overall video structure. Video analysis major components are video scene segmentation, shot boundary detection, key frame selection, and extraction [6–8]. The main use of key frame extraction is to reduce the redundant

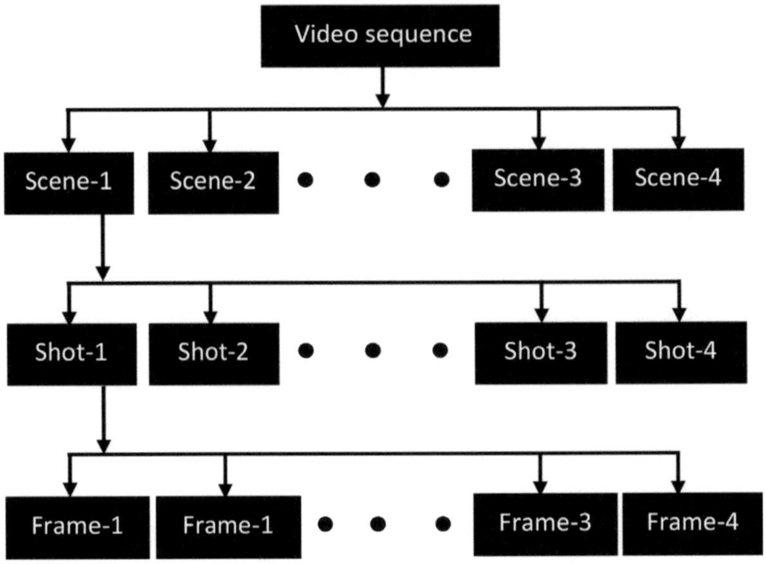

Figure 1.
Structure of video.

frames in a video to make a video scene readable and compact and prepare video sequences for faster processing.

Conventional key frame extraction methods eliminate the redundant and similar frame in a video without affecting the semantic details visual content. These techniques inputs are either a complete video or a video is divided into a set of shots by shot boundary detection methods. As shown in **Figure 1**, the shot is a consecutive, adjacent sequence of frames captured by the video camera. Thus, in this chapter we propose an efficient approach for video key frame extraction, which is faster, accurate, and computationally efficient. This chapter is organized into the following sections. Section 1 gives an introductory part of video structure and the importance of key frame extraction in a video surveillance system. Section 2 describes the existing approach for key frame size selection algorithms. Section 3 describes the existing key frame extraction methods with its issues and challenges. Section 4 describes the proposed approach for key frame extraction. Section 5 discusses experimental results and possible future directions. Finally, the chapter concluded with a discussion in Section 6.

2. Key frame size estimation methods available in the literature

The major problem we face in a key frame extraction algorithm is computing the size or number of key frames for a specific video sequence. In literature, there are several methods available for the key frame size estimation. In this section, we have discussed these methods in brief. In [3] approach the author has considered one key frame for each shot of a video. The selection of the key frame in each shot is based on the maximum entropy value of each shot. This consideration is not appropriate and accurate for the video which is having a big shot. Again, many of the useful frames of the video are discarded due to the pre-defined fixed selection of key frames. Lesser key frame extraction does not solve the problem. A set of key frames having necessary and sufficient representation of the visual content of the video is required in the output. In other proposed approaches, first, middle, or ending location frames in the shot are considered. But the resulting frames are having a low correlation with each other in visual content. These methods are computationally less complex, but having less accuracy. In [9, 10], authors have described the three different ways of identifying the key frames in a video sequence. Each method is described in brief as follows.

2.1 Priori knowledge base as a fixed number

In this method, a pre-defined number of fixed key frames are considered as a fixed value before the key frame extraction process. Consider "k" as the number of key frames, and then the key frame set K_r is defined by Eq. (1):

$$K_r = \{F_{i1}, F_{i2}, F_{i3}, \dots ..F_{ik}\} \tag{1}$$

The sequence of video frames is the change as per the type of video. The specific summarization of key frames is defined by Eq. (2):

$$\{K_{f1}, K_{f2}, K_{f3}, \dots .K_{fk}\} = \sum_{n}^{1} min_{ri}\{Dist(K_r, V, \delta)\} \tag{2}$$

where,

$1 \leq r_i \geq n$ and

n is represented as a number of frames in a video, δ represents the key frame summarization factor, and *Dist* represents the distance measure, i.e., used for computing dissimilarity between frames. The δ in this method is useful for maintaining a lesser number of key frames by covering complete visual content details in the video.

2.2 Posteriori knowledge base as unknown

In this method, the number of key frames is not fixed. The number of key frames is unknown until the key frame extraction process gets completed. The key frame size is depending upon the type of content of the video frame. If the video scene consists of dynamic action movements, then the number of key frames is more otherwise less for static video scenes. Key frame generation can be represented by Eq. (3):

$$\{K_{f1}, K_{f2}, K_{f3}, \dots .K_{fk}\} = \sum_{n}^{1} min_{ri}\{K|Dist(K_r, V, \gamma)\} \tag{3}$$

where γ parameter is used for tolerance to dissimilarity level. Another parameter is similar to the previous method.

2.3 Determined-fixed number

In this method, the number of key frames is predetermined before the whole process key frame extraction process. In [11, 12] approaches key frames are extracted using the clustering technique. The key frame extraction algorithms stop when extracted key frame size matched with the pre-defined key frame value.

3. Key frame extraction methods with its issues and challenges

In literature, there are several methods to extract key frames. Hannane et al. [13] and Hu et al. [14] categorize the key frame extraction into different categories as a sequential comparison of frames, global comparison of frames, the minimum correlation between frames, minimum reconstruction error in frames, temporal variance between frames, maximum coverage of video frames, reference key frame, curve simplification, key frame extraction using clustering, object- and event-based key frame extraction, and panoramic key frames. Each of these methods is described in brief as follows.

3.1 Sequential comparison of frames

In this method, each frame of a video sequence is compared with the previously extracted key frame. If the difference between the extracted key frame and the current key frame is high, then this frame is considered as the new key frame. In [13] key frames are extracted based on the color histogram comparison between the current and previous frames of a video sequence. The main advantage of this method is that it is simple and computationally less complex. But the disadvantage is that the extracted key frame consists of redundant key frames.

3.2 Universal frame comparison

In this method, the global difference between frames in a shot is computed using a predetermined objective function, which is application-specific. Zhuang et al. in [9] describes the different objective functions for comparison of frames in the shot. Each of these functions is discussed in brief as follows.

3.3 Minimum associations

In this method, relevant key frames are generated from a shot by reducing the summation of the association between frames. The extracted key frames are tightly coupled with each other. Liu in [3] uses a graph-based approach to extract distinct key frames with their association. Weight directed graph is used to represent the shot, and the shortest path is computed using the A* algorithm. The frames, which are having minimum association and less correlation, are represented as key frames in the shot.

3.4 Minimum reformation error

In this method, the key frames are extracted by reducing the variation between the prevision frame and set of frames in a shot. The prevision frame is generated by the numeric analysis method interpolation. Chao et al. in [15] presented an approach to select a pre-defined set of key frames and reduce the frame reformation error. In [16] a combined approach of the prevision frame-based approach and a pre-defined set of key frame selection approach is proposed. This method uses the motion-based features.

3.5 Similar temporal variance

In these methods, frames having similar variance are selected as the key frames of the specific shot [17]. The sum of temporal variance between all frames is selected as an objective function. The temporal variance is computed by the summation of change in the frame content in a shot.

3.6 Maximum key frame representation coverage

In this method, the representation coverage of a key frame means a number of frames in a shot that a key frame can cover [18]. This method can be useful in the size of the key frame selection. The advantage of this method over a universal comparison method is that the extracted key frames are maintainable and consist of global context information of a shot. The only disadvantage of this method is that it is computationally complex.

3.7 Predetermined reference frame

In this method, a key frame is generated by comparing the predetermined reference frame and each frame in a shot [19]. The main advantage of this method is that it is not computationally complex and easy to implement. Its drawback is that it does not represent the global context in a shot efficiently.

3.8 Trajectory curve simplification

In this method, the trajectory curve is generated from the frames. The curve consists of a sequential combination of points in the feature. Calic and Izquierdo in [20] presents a dynamic method for change detection in the scene and the key frame generation. The frame difference metric is computed using the small size block features in a scene. After that contour detection method is used for trajectory curve plotting using the metric.

3.9 Cluster-based key frame extraction

In this method, key frame clusters are created using the data points and features of video sequences. The set of key frames is created with frames that have the closest distance from the center of the cluster. In [21, 22] fuzzy K-means- and fuzzy C-means-based methods for the key frame selection are presented. The clusters are generated based on the different features like motion sequences and the distance matrix score. In [23] an approach that combined K-means and mean squared error for the key frame selection is presented. Pan et al. in [24] proposed an enhanced fuzzy C-means clustering algorithm for the key frame selection. Clusters are generated using the color feature. The key frames having the highest entropy are considered as a key frame from each cluster. The advantage of cluster-based approaches is that it covers the global characteristics of the scene. The disadvantage of these methods is that it requires a high computational cost for cluster generation and feature extraction from the scene.

3.10 Event-driven key frame extraction

In this method, the extracted key frame consists of event and object details. The advantage of this method is that each key frame describes the object motion pattern, object, and event details [25]. The drawback of this method is that the predefined rules need to be defined as per the application, identifying objects and events in a key frame. Hence, the accuracy of this algorithm depends upon the preassumption parameters set before the key frame extraction algorithm is executed.

3.11 Full details key frame extraction (panoramic frame)

In this method, the key frame consists of the complete detail of a scene in a shot. Papageorgiou and Poggio in [25] presented a key frame extraction approach using

Method name	Characteristics	Advantage	Shortcomings of the method	Ref.	Year
Clustering method (Zhuang et al.)	Analysis of short boundary video	Faster processing	• Less key frame selection for single-shot activity • More key frame selection for multiple	[9]	1998
Entropy method (Mentzelopoulos et al.)	Best method for unpredictable dataset	Local feature selection	• External effects such as lighting condition affect the performance	[10]	2012
Histogram method (Rasheed et al.)	Similarity measure between key frames	High-level segmentations	• Cannot consider the local similarities	[11]	2015
Motion analysis method (Wolf et al.)	Optical flow-based analysis	Faster mid-range key frame selection	• Highly depends on the static frame references	[12]	2016
Triangle-based method (Liu et al.)	Determination of the motion characteristics	Reduces the motion effects on the video	• Cannot detect the color-based information change	[3]	2016
3D augmentation method (Chao et al.)	Processing short and fast motion video data	Combines the video data into multidimensional model	• Highly time complex	[15]	2018
Optimal key frame selection method (Sze et al.)	Best method for continuously growing video sequence by adopting the temporary key frame	Faster processing due to probabilistic analysis	• Highly time complex	[16]	2017
Context-based method (Chang et al.)	Best method for repetitive information contents	Generates a multilevel abstract of the information	• Information loss due to less key frame selection	[17]	2017
Motion-based extraction method (Luo et al.)	Adopts the advantages from digital capture devices	Reduces the spatiotemporal effects	• High-quality video information expected	[18]	2015
Robust principal component analysis method (Dang et al.)	Adopts the decomposition method for sparse component analysis	Analyzes the frames for consumer videoswith fewer contents or rapid content shift	• Assumptions are not always reflecting better results	[19]	2010

Table 1.
Recently used pedestrian databases by the researchers.

the homography matrix. The main advantage of this method is that it covers the global context of the shot. The drawback of this method is that it is having high computational complexity. The comparative analysis of recently utilized key frame extraction algorithms is shown in **Table 1**. The comparison is performed in terms of characteristics, advantages, and shortcomings of the method.

4. Proposed methodology for key frame extraction

The proposed approach is based on the combination of the histogram and deep learning to extract the relevant key frame from the video sequence. **Figure 2** shows the main steps of the proposed framework. The steps of key frame extraction include (1) video reading from the database, (2) frame extraction from video, (3) preprocessing, (4) histogram generation, (5) comparison of the histogram, (6) distinct key frame generation, and (7) key frame extraction using convolution neural network (CNN). Each of these steps is described in subsequent subsections.

4.1 Video reading from database

We have tested this algorithm on the various publicly available datasets and on our own behavioral dataset. The first step is to read a video from the database. The raw video sequence selected from database is represented by Eq. (4):

$$V_i = \{V_1, V_2, V_3, \dots .V_k\} \tag{4}$$

where $1 \geq V \leq k$.

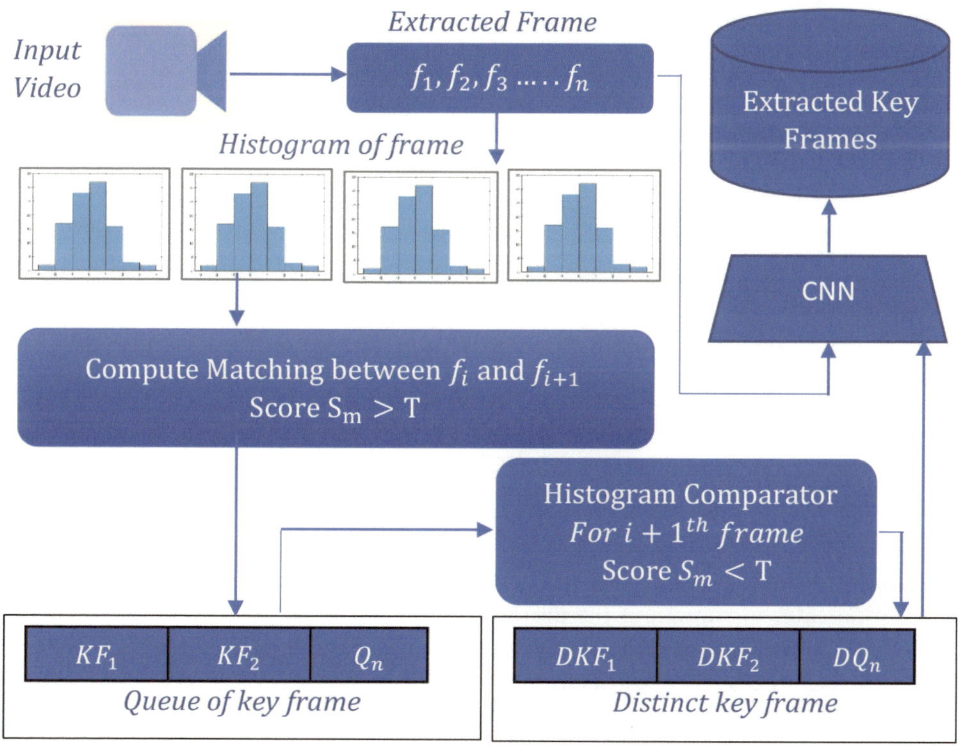

Figure 2.
Proposed framework for key frame extraction.

4.2 Frame extraction from video

The number of frames is extracted from the video selected in step 1. The extracted frames are stored in a local directory for further processing. It is represented by Eq. (5):

$$F_i = \{F_1, F_2, F_3, \ldots .F_n\} \tag{5}$$

where $1 \geq F \leq n$.

4.3 Preprocessing of frames

In the preprocessing step, the key frame queue initialized with $Q_k = 0$. The key frame queue Q_k initialized with zero because in the initial step key frame is zero. Next, the extracted frames of step 2 are converted from RGB model space to the HSV model space. This conversion is necessary to get a more specific color, gray shade, and brightness information. In HSV model space, hue is the color portion of the model, expressed as a number from 0 to 360. Saturation describes the amount of gray in a particular color, ranging from 0 to 100%. The value component represents the intensity of the color, ranging from 0 to 100%, where 0 is completely black and 100 is the brightest and reveals the most color.

4.4 Histogram generation

In this step, the normalized histogram is generated from the hue-saturation and value component in order to compare the adjacent frame. The normalized histogram is generated for contrast enhancement and compact representation of intensity and color information of the frame. Normalized histogram H_n is computed by Eq. (6)

$$H_n = \frac{number\ of\ pixels\ with\ intensity\ n}{total\ number\ of\ pixels} \tag{6}$$

where, n indicates possible intensity value.

4.5 Histogram comparison

In this step, the normalized histogram H_n is generated for each frame, and adjacent frame histogram is compared using the Bhattacharyya distance measure. It is defined by Eq. (7):

$$d(Hn_1, Hn_2) = \sqrt{1 - \frac{1}{\sqrt{Hn_1 Hn_2 N^2}} \sum_I \sqrt{Hn_1(F_p).Hn_2(F_c)}} \tag{7}$$

where:
Hn_1 indicates histogram of the previous frame $F_{p.}$
Hn_1 indicates histogram of the current frame F_c.
N indicates the number of histogram bins

The Bhattacharyya distance $d(Hn_1, Hn_2)$ is the result of a comparison of the matched score (S_m). The S_m value ranges from 0 to 1. The value 0 indicates an exact match of the content of the video frame, 0.5 is half match and 1 represents mismatch. Next, different conditions are checked to match to extract dissimilar

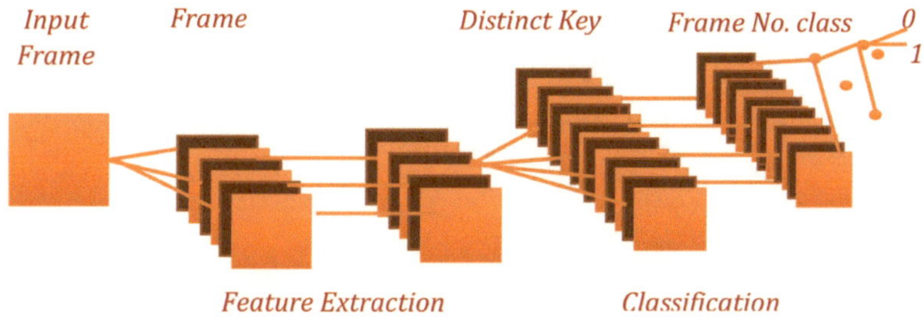

Figure 3.
A CNN for proposed key frame extraction algorithm.

frames and similar frames. The different conditions of the score (S_m) are compared with a threshold (T) as:

1. *if* $(S_m > T)$ *then* current frame is dissimilar than the previous
2. *if* $(Q_k! = 0)$ then
3. Add a frame in the queue of key frame $Q_k \leftarrow f_i$

4.6 Distinct key frame generation

In this step, the distinct key frame is selected, and redundant key frames are removed from the Frame queue as follows:

1. for each frame FQ_i in Q_k
2. $S_m \leftarrow (FQ_i, F_i)$
3. if $(S_m < T)$ then current frame is dissimilar than the previous.
4. Add a frame in the queue of key frame $DQ_k \leftarrow f_i$.

4.7 Key frame extraction using a convolution neural network

CNN is composed of two basic parts of feature extraction and classification. Feature extraction includes several convolution layers followed by max-pooling and an activation function. The classifier usually consists of fully connected layers as shown in **Figure 3**.

Extracted distinct key frames are used as testing queries in classification phase, and input frames features are extracted using the CNN feature extraction module, and learn features are matched with distinct key frame features to obtain the best match frame which is considered as key in the output as a frame index number. The key frame extraction and CNN approach perform in parallel to obtain efficiency.

5. Experiment results and discussion

In this section, we have evaluated the efficiency of the proposed method on a publicly available database and our own human action database.

Data source	Purpose	# Image or video clips	Annotation	Environment	Ref.	Year
MIT	City street pedestrian segmentation, detection, and tracking	709 pedestrian images, 509 training and 200 test images	No annotated pedestrian	Daylight scenario	[26] [27]	2000, 2005
Caltech Pedestrian dataset	Detection and tracking of pedestrian walking on the street	250,000 frames (in 137 approximately minute-long segments)	350,000 bounding boxes and 2300 unique pedestrians were annotated	Urban environment	[27]	2012
GM-ATCI	Rear-view pedestrian segmentation, detection, and tracking	250 video sequences	200 K annotated pedestrian bounding boxes	Dataset was collected in both day and night scenarios, with different weather and lighting conditions	[28]	2015
Daimler	Detection and tracking of pedestrian	15,560 pedestrian samples, 6744 negative samples	2D bounding box overlap criterion and float disparity map and a ground truth shape image	Urban environment	[29]	2016
NICTA 2016	Segmentation, pose estimation, learning of pedestrian	25,551 unique pedestrians, 50,000 images	2D ground truth image	Urban environment	[30]	2016
MS COCO 2018	Object detection, segmentation, key point detection, DensePose detection	300,000, 2 million instances, 80 object categories	5 captions per image	Urban environment	[31]	2018
Mapillary Vistas dataset 2017	Semantic understanding street scenes	25,000 images, 152 object categories	Pixel-accurate and instance-specific human annotations for understanding street scenes	Urban environment	[32]	2017
MS COCO 2017	Recognition, segmentation, captioning	328,124 images, 1.5 million object instances	Segmented people and objects	Urban environment	[33]	2017
MS COCO 2015	Recognition, segmentation, captioning	328,124 images, 80 object categories	Segmented people and objects	Urban environment	[34]	2015
ETH	Segmentation, detection, tracking	Videos	The dataset consists of other traffic agents such as different cars and pedestrians	Urban environment	[35]	2010
TUD-Brussels	Detection, tracking	1092 image pairs	1776 annotated pedestrian	Urban environment	[33]	2009

Data source	Purpose	# Image or video clips	Annotation	Environment	Ref.	Year
INRIA	Detection, segmentation	498 images	Annotations are marked manually	Urban environment	[34]	2005
CVC-ADAS	Detection, tracking	60,000 frames	7900 annotated pedestrians	Urban environment	[35]	2009
PASCAL VOC 2012	Detection, classification, segmentation	11,530 images, 20 object classes	27,450 ROI annotated 6929 segmentations	Urban environment	[36]	2012
Pedestrian behavior dataset (own DB)	Pedestrian behavior recorded in the college environment	50 human behavior datasets	No annotated pedestrian	Daylight scenario	—	—

Table 2.
Pedestrian databases used for the experiment for key frame extraction.

Type of features	Recall	Precision	CPU time (ms)
Proposed key frame extraction algorithm	0.95	0.92	0.50
Discrete cosine coefficients and rough sets theory based [1]	0.88	0.82	0.90
Content relative thresholding technique based [2]	0.80	0.81	0.80
Multi-scale color contrast, relative motion intensity, and relative motion consistency based [3]	0.83	0.80	0.90
Color and structure feature based [4]	0.80	0.86	0.98

Table 3.
Comparative analysis of mean, recall, and precision and CPU time achieved by different techniques.

The results demonstrate significant improvement over the conventional methods and with low time complexity. Next, in subsequent sections, the various experiments conducted are discussed as follows.

5.1 Dataset analysis

The performance of a key frame extraction technique was evaluated and compared with the state-of-the-art methods using benchmark databases. We have taken sample videos of benchmark database and human action database as shown in **Table 2**.

5.2 Computational complexity of the proposed system

The proposed methodology is clearly superior to the rest of the techniques for key frame extraction as shown in **Table 3**. The comparative analysis of recall and precision metric for each video sequence is shown in **Figure 4**. It is observed that the proposed approach of key frame extraction achieves the highest values for recall and precision for all the video sequences. A maximum value of one of the metrics is generally not sufficient. The precision metric is used to measure the ability of a technique to retrieve the most precise results. A high value of precision means better relevance between the key frames. However, a high value of precision can be

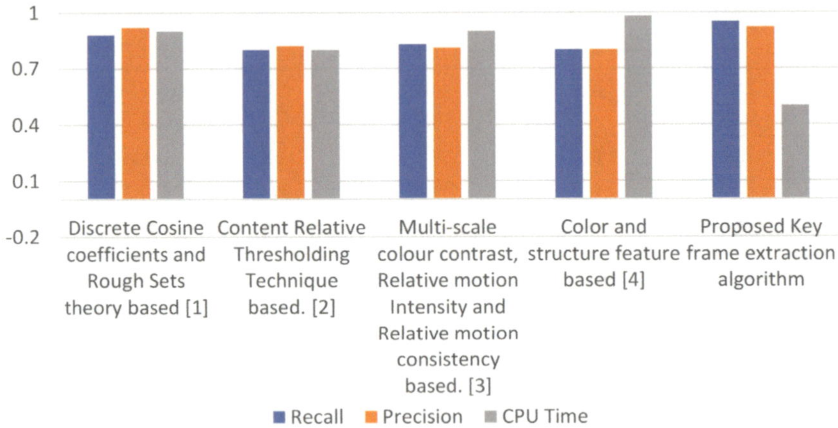

Figure 4.
Recall (R), precision (P), and computational time achieved by different techniques on video dataset of ***Table*** *2.*

Figure 5.
Input video frame.

Figure 6.
Extracted key frame from video.

achieved by selecting very few key frames in a video sequence. The speed and accuracy of both parameters are important in the key frame extraction algorithm. If the algorithm is slow, then the throughput of the system gets affected. It is also necessary that extracted key frames are relevant and accurate. Further, it will affect the other process, such as object detection, classification, object description, etc., respectively (**Figures 5** and **6**).

5.3 Qualitative result of frame extraction

Qualitative results from the proposed deep learning approach for the key frames extraction algorithm are shown in **Figure 7**. The figure illustrates the relevant and non-redundant key frames are extracted from the video sequence. The dataset consists of 7 suspicious student behavior. The pedestrian behaviors are recorded at prominent places of the college in different academic activities.

Figure 7.
Qualitative results of the proposed key frame extraction method on a sample video of student pedestrian dataset. (a) Frames extracted from sample video of dataset (First three-column). (b) Key frames extracted from sample video (Forth column).

6. Conclusions

This chapter describes and evaluates the methodologies, strategies, and stages involved in video key frame extraction. It also analyzes the issue and challenges of each of the key frame extraction methods. Based on the literature survey, most of the available techniques proposed by the earlier researchers can perform key frame extraction. However, most of them failed to encounter the trade-off problem between accuracy and speed. The proposed framework and approach give significant improvements for key frame extraction irrespective of the video length rather on the content type. This is made possible due to the histogram-based comparison of video scene content and convolution neural network-based deep features approach. With significantly satisfactory results, this work can generate a key frame dynamically from any video sequence. We have performed experiments on the publicly available database and obtained encouraging results.

Author details

Ujwalla Gawande[1]*, Kamal Hajari[1] and Yogesh Golhar[2]

1 Department of Information Technology, Yeshwantrao Chavan College of Engineering, Wanadongri, Maharashtra, India

2 Department of Computer Science and Engineering, G.H. Raisoni Institute of Engineering and Technology, Nagpur, Maharashtra, India

*Address all correspondence to: ujwallgawande@yahoo.co.in

References

[1] Aigrain P, Zhang H, Petkovic D. Content-based representation and retrieval of visual media: A state-of-the-art review. Multimedia Tools and Applications. 1996;**3**(3):179-202

[2] HongJiang Z, Wang JYA, Altunbasak Y. Content-based video retrieval and compression: A unified solution. In: Proceedings of IEEE International Conference on Image Processing (ICIP); October 26–29, 1997; Santa Barbara, CA. pp. 13-16

[3] Liu T, Zhang H-J, Qi F. A novel video key-frame-extraction algorithm based on perceived motion energy model. IEEE Transactions on Circuits and Systems for Video Technology (TCSVT). 2003;**13**(10):1006-1013

[4] Gargi U, Kasturi R, Strayer SH. Performance characterization of video-shot-change detection methods. IEEE Transactions on Circuits and Systems for Video Technology (TCSVT). 2000; **10**(1):1-13

[5] Liu G, Zhao J. Key frame extraction from MPEG video stream. In: Second symposium International Computer Science and Computational Technology (ISCSCT'09); December 26–28, 2009; Huangshan, P.R. China. pp. 007-011

[6] Gawande U, Golhar Y. Biometric security system: A rigorous review of unimodal and multimodal biometrics techniques. International Journal of Biometrics (IJBM). 2018;**10**(2):142-175

[7] Gawande U, Golhar Y, Hajari K. Biometric-based security system: Issues and challenges. In: Intelligent Techniques in Signal Processing for Multimedia Security. Studies in Computational Intelligence, Vol. 660; October, 2017; Cham: Springer. pp. 151-176

[8] Gawande U, Zaveri M, Kapur A. A novel algorithm for feature level fusion using SVM classifier for multibiometrics-based person identification. Applied Computational Intelligence and Soft Computing. 2013; **2013**(9):1-11

[9] Zhuang Y, Rui Y, Huang TS, Mehrotra S. Adaptive key frame extraction using unsupervised clustering. In: Proceedings of IEEE International Conference on Image Processing (ICIP); October 7, 1998; Chicago, IL, USA, pp. 866-870

[10] Mentzelopoulos M, Psarrou A. Key frame extraction algorithm using entropy difference. In: Proceedings of the 6th ACM SIGMM International Workshop on Multimedia Information Retrieval, MIR; 15–16 October, 2004; New York, NY, USA. pp. 39-45

[11] Rasheed Z, Shah M. Detection and representation of scenes videos. IEEE Transactions on Multimedia. 2005;7(6): 1097-1105

[12] Wolf W. Key frame selection by motion analysis. In: Proceedings of IEEE International Conference on Acoustics, Speech Signal Processing; May 9, 1996; Atlanta, GA, USA, pp. 1228-1231

[13] Hannane R, Elboushaki A, Afdel K, Naghabhushan P, Javed M. An efficient method for video shot boundary detection and keyframe extraction using SIFT-point distribution histogram. International Journal of Multimedia Information Retrieval. 2016;**5**(2):89-104

[14] Weiming H, Xie N, Li L, Zeng X, Maybank S. A survey on visual content-based video indexing and retrieval. IEEE Transactions on Systems, Man, and Cybernetics, Part C (Applications and Reviews). 2011;**41**(6):797-819

[15] Chao G, Tsai Y, Jeng S. Augmented 3-D keyframe extraction for surveillance videos. IEEE Transactions

on Circuits and Systems for Video Technology (TCSVT). 2010;**20**(11): 1395-1408

[16] Sze KW, Lam K-M, Qiu G. A new key frame representation for video segment retrieval. IEEE Transactions on Circuits and Systems for Video Technology (TCSVT). 2005;**15**(9): 1148-1155

[17] Chang HS, Sull S, Lee SU. Efficient video indexing scheme for content-based retrieval. IEEE Transactions on Circuits and Systems for Video Technology (TCSVT). 1999;**9**(8): 1269-1279

[18] Luo J, Papin C, Costello K. Towards extracting semantically meaningful key frames from personal video clips: From humans to computers. IEEE Transactions on Circuits and Systems for Video Technology (TCSVT). 2009; **19**(2):289-301

[19] Dang C, Radha H. RPCA-KFE: Key frame extraction for video using robust principal component analysis. IEEE Transactions on Image Processing (TIP). 2015;**24**(11):3742-3753

[20] Calic J, Izquierdo E. Efficient key-frame extraction and video analysis. In: Proceedings of International Conference on Information Technology: Coding and Computing; April, 2002; Las Vegas, NV, USA, pp. 28-33

[21] Nasreen A, Roy K, Roy K, Shobha G. Key frame extraction and foreground modelling using K-means clustering. In: International Conference on Computational Intelligence, Communication Systems and Networks (CICSYN); Latvia; 2015. pp. 141-145

[22] Yu XD, Wang L, Tian Q, Xue P. Multilevel video representation with application to keyframe extraction. In: Proceedings Multimedia Modelling Conference, Australia; 2004. pp. 117-123

[23] Zhang Q, Yu S-P, Zhou D-S, Wei X-P. An efficient method of key-frame extraction based on a cluster algorithm. Journal of Human Kinetics. 2013;**39**(1): 5-13

[24] Pan R, Tian Y, Wang Z. Key-frame extraction based on clustering. In: Proceedings of IEEE Progress in Informatics and Computing; December, 2010; China. pp. 867-871

[25] Papageorgiou C, Poggio T. A trainable system for object detection. International Journal of Computer Vision. 2000;**38**(1):15-33

[26] Dollar P, Wojek C, Schiele B, Perona P. Pedestrian detection: An evaluation of the state of the art. IEEE Transactions on Pattern Analysis and Machine Intelligence. 2012;**34**(4): 743-761

[27] MIT Pedestrian Dataset. Center for Biological and Computational Learning at MIT and MIT. 2005. Available from: http://cbcl.mit.edu/software-datasets/PedestrianData.html [Accessed: 22 September 2018]

[28] Levi D, Silberstei S. Tracking and motion cues for rear-view pedestrian detection. In: 18th IEEE Intelligent Transportation Systems Conference (ITSC); September 15–16, 2015; Spain. pp. 664-671

[29] Li X, Flohr F, Yang Y, Xiong H, Braun M, Pan S, et al. A new benchmark for vision-based cyclist detection. In: IEEE Intelligent Vehicles Symposium (IV), Gothenburg, Sweden, June 19–22, 2016. pp.1028-1033

[30] Campbell D, Petersson L. GOGMA: Globally-Optimal Gaussian Mixture Alignment. In: Proceedings of the 2016 IEEE Conference on Computer Vision and Pattern Recognition (CVPR), Las Vegas, USA, IEEE; June, 2016

[31] Pellegrini S, Ess A, Van Gool L. Wrong turn – no dead end: A stochastic

pedestrian motion model. In: International Workshop on Socially Intelligent Surveillance and Monitoring (SISM'10), in Conjunction with International Conference on Computer Vision and Pattern Recognition (CVPR); June 13–18, 2010; San Francisco, CA, USA

[32] Wojek C, Walk S, Schiele B. Multi-cue onboard pedestrian detection. In: IEEE Conference on Computer Vision and Pattern Recognition (CVPR); June 20–25, 2009; Miami, Florida, USA

[33] Dalal N. Finding people in images and videos. GRAVIR - IMAG - Graphisme, Vision et Robotique, Inria Grenoble - Rhône-Alpes, CNRS - Centre National de la Recherche Scientifique [PhD thesis]; 2006

[34] Everingham M, Eslami SMA, Van Gool L, Williams CKI, Winn J, Zisserman A. The Pascal visual object classes challenge: A retrospective. International Journal of Computer Vision. 2015;**111**(1):98-136

[35] CVC-ADAS Pedestrian dataset. 2012. Available from: http://adas.cvc.uab.es/site/ [Accessed: 22 September 2018]

[36] Lin T-Y, Maire M, Belongie S, Hays J, Perona P, Ramanan D, et al. Microsoft coco: Common objects in context. In: European Conference on Computer Vision; Febrauary, 2015 Springer; 2014. pp. 740-755

Chapter 8

Machine Translation and the Evaluation of Its Quality

Mirjam Sepesy Maučec and Gregor Donaj

Abstract

Machine translation has already become part of our everyday life. This chapter gives an overview of machine translation approaches. Statistical machine translation was a dominant approach over the past 20 years. It brought many cases of practical use. It is described in more detail in this chapter. Statistical machine translation is not equally successful for all language pairs. Highly inflectional languages are hard to process, especially as target languages. As statistical machine translation has almost reached the limits of its capacity, neural machine translation is becoming the technology of the future. This chapter also describes the evaluation of machine translation quality. It covers manual and automatic evaluations. Traditional and recently proposed metrics for automatic machine translation evaluation are described. Human translation still provides the best translation quality, but it is, in general, time-consuming and expensive. Integration of human and machine translation is a promising workflow for the future. Machine translation will not replace human translation, but it can serve as a tool to increase productivity in the translation process.

Keywords: machine translation, statistical machine translation, neural machine translation, evaluation, post-editing

1. Introduction

Machine translation (MT) investigates the approaches to translating text from one natural language to another. It is a subfield of computational linguistics that draws ideas from linguistics, computer science, information theory, artificial intelligence, and statistics. For a long time, it had a bad reputation because it was perceived as low quality. Especially in the last two decades, we have been witnessing great progress in MT quality, which made it interesting also for the use in the translation industry. Its quality is still lower than human translation, but that does not mean it does not have good practical uses. In the past, translation agencies and other professional translators were the only actors in the translation industry, but, in recent years, we have been faced with the rapidly growing range of machine translation solutions entering the market and being of practical use. There is increasing pressure on the translation industry in terms of price, volume, and turnaround time. The emergence of commercial applications for MT is a welcome change in translation processes. In professional or official circumstances, human translation is inevitable, as humans are essential to making sure a translation is grammatically correct and carries the same meaning as the original text.

Machine translation is appropriate in different circumstances, mainly for unofficial purposes or for providing content for a human translator to improve upon it. MT has proved to be able to speed up the whole translation process, but it cannot replace the human translator. The questions that researchers in the translation industry are trying to answer today are: How much can human translators benefit from using MT? How could MT be integrated efficiently into translation processes? If MT is integrated into the translation workflow, will the quality of translation remain at the same level? These questions will not be answered explicitly in this chapter, but an effort will be made to show that MT is worth being part of the translation process, as its quality can be evaluated reliably. MT opens new opportunities for translators through using MT output only as a suggestion and, if necessary, post-editing it to the desired quality. It could be much faster than translation from scratch. This process is further discussed in the penultimate section of the chapter.

The aim of this chapter is to overview the methods of machine translation and the methods of the evaluation of its quality. This chapter is organised as follows. In Section 2, different approaches for machine translation are described: rule-based MT in Section 2.1, example-based MT in Section 2.2, statistical MT in Section 2.3, hybrid MT in Section 2.4, and neural MT in Section 2.5. Not all languages are equally difficult for MT. Section 3 exposes common problems when dealing with morphologically rich languages. Sections 4–7 are devoted to MT evaluation. In Section 5, basic metrics for automatic MT evaluation are described and in Section 6, the more advanced ones. Automatic MT evaluation makes sense if it gives similar results as manual evaluation. Section 7 discusses how the correlation between automatic and manual MT evaluation is determined. MT is never perfect. Section 8 discusses post-editing MT to correct the mistakes and make MT of practical use. Section 9 concludes this chapter.

2. Machine translation

Computer scientists began trying to solve the problem of MT in the 1950s. The first published machine translation experiment was performed by the Georgetown University and IBM. It involved automatic translation of more than 60 Russian sentences into English. The system had only 6 grammar rules and 250 lexical items in its vocabulary. It was by no means a fully featured system. The sentences for translation were selected carefully, as the idea of the experiment was to attract governmental and public interest and funding by showing the possibilities of MT. Many problems of MT had come to light right after, and, consequently, for a long time, MT was present only as a research area in computational linguistics. Overtime, different approaches for MT were defined and gained maturity for practical use today. The history of the development of MT approaches is given in **Figure 1**. In [1], it has been shown that 22% of the MT users in the translation industry use rule-based MT systems, 50% use statistical MT systems, and 36% of them use hybrid MT systems.

2.1 Rule-based machine translation

The first approaches for MT were based on linguistic rules that were used to parse the source sentence and create the intermediate representation, from which the target language sentence was created. Such approaches are appropriate to translate between closely related languages. The rule-based machine translation methods include dictionary-based MT, transfer-based MT, and interlingual MT.

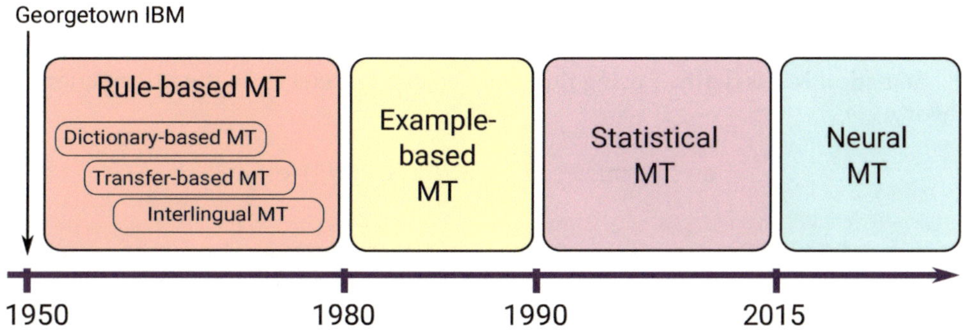

Figure 1.
Timeline of MT evolution.

Dictionary-based MT uses entries in a language dictionary to find a words equivalent in the target language. Using a dictionary as the sole information source for translation means that the words will be translated as they are translated in a dictionary. As this is, in many cases, not correct, grammatical rules are applied afterwards.

Transfer-based MT belongs to the next generation of machine translation. The source sentence is transformed into an intermediate, less language-specific structure. This structure is then transferred into a similar structure of the target language, and, finally, the sentence is generated in the target language. The transfer uses morphological, syntactic, and/or semantic information about the source and target languages.

In interlingual MT, the source sentence is transformed into an intermediate, artificial language. It is a neutral representation that is independent of any language. The target sentence is then generated out of the interlingua.

To be useful in practice, rule-based MT systems consist of large collections of rules, developed manually over time by translation experts, mapping structures from the source language to the target language. They are costly and time-consuming to implement and maintain. As rules are added and updated, there is the potential of generating ambiguity and translation degradation. Rule-based MT requires linguistic experts to apply language rules to the system.

2.2 Example-based machine translation

Example-based MT is based on the idea of analogy. It is grounded upon a search for analogous examples of sentence pairs in the source and target languages. Example-based MT belongs to corpus-based approaches because examples are extracted from large collections of bilingual corpora. Given the source sentence, sentences with similar sub-sentential components are extracted from the source side of the bilingual corpus, and their translations to the target language are then used to construct the complete translation of the sentence.

2.3 Statistical machine translation

Statistical MT is based on statistical methods [2]. It also belongs to corpus-based approaches, as statistical methods are applied on large bilingual corpora. Building a statistical MT system does not require linguistic knowledge. Statistical MT utilises statistical models generated from the analysis of texts, being either monolingual or bilingual. It is called training data. If more training data are available, better and larger MT systems can be built. Statistical MT systems are computationally expensive to build and store. Statistical MT can be adapted easily to a specific domain if enough bilingual and/or monolingual data from that domain are available.

Statistical MT is defined using the noisy-channel model from the information theory:

$$\mathbf{e} = \arg\max_{\mathbf{e}} P(\mathbf{e}|\mathbf{f}) = \arg\max_{\mathbf{e}} P(\mathbf{e})P(\mathbf{f}|\mathbf{e}). \tag{1}$$

where **f** is the source sentence and **e** is the target sentence. The source sentence consists of words f_j and the target sentence of words e_i. Words f_j belong to the source vocabulary F and the words e_i to the target vocabulary E. In the phrase-based model, the source sentence f is broken down into I phrases $\overline{f}_i$, and each source phrase $\overline{f}_i$ is translated into a target phrase $\overline{e}_i$.

Standard phrase-based SMT models consist of three components:

1. A translation model of phrases (denoted as $\phi(\overline{f}|\overline{e})$). In practice, both translation directions, with the proper weight setting, are used: $\phi(\overline{f}|\overline{e})$ and $\phi(\overline{e}|\overline{f})$.

2. A reordering model (denoted as d). It is based on distance. The reordering distance is computed as $start_i - end_{i-1} - 1$, where $start_i$ is the position of the first word in phrase i, end_{i-1} is the position of the last word of phrase $i-1$, and d is the probability distribution of reordering.

3. A language model (denoted as $p_{LM}(\mathbf{e})$). It makes the output a fluent sequence of words in the target language and is most commonly an n-gram language model.

Log-linear models of phrase-based SMT are most commonly used:

$$p(\mathbf{e}, a|f) = \exp\left[\lambda_\phi \sum_{i=1}^{I} \log\phi\left(\overline{f}_i|\overline{e}_i\right) + \lambda_d \sum_{i=1}^{I} \log d(start_i - end_{i-1} - 1) + \lambda_{LM} \sum_{i=1}^{N} \log p_{LM}(e_i|e_1 \dots e_{i-1})\right]. \tag{2}$$

where a is an alignment between source and target sentences and N is the length of the target sentence.

Statistical MT faces many obstacles. Data sparsity of highly inflected languages limits the effectiveness of statistical MT. Advanced statistical MT systems try to overcome the limitations by introducing data preprocessing and data post-processing. In **Figure 2** data preprocessing is used, where morphosyntactic tags (MSD) and lemmas are assigned to words and used in translation and language models. Reordering model captures short-term dependencies. The operation sequence model (OSM) is able to capture long-distance dependencies [3]. It models translation by a linear sequence of operations. The operation generates translation, performs reordering, jumps forward and backward, etc. Having morphosyntactic tags and lemmas available, OSM could be constructed based on them, as depicted in **Figure 2**.

2.4 Hybrid machine translation

While statistical methods still dominate research work in MT, most commercial MT systems were, from the beginning, only rule-based. Recently, boundaries

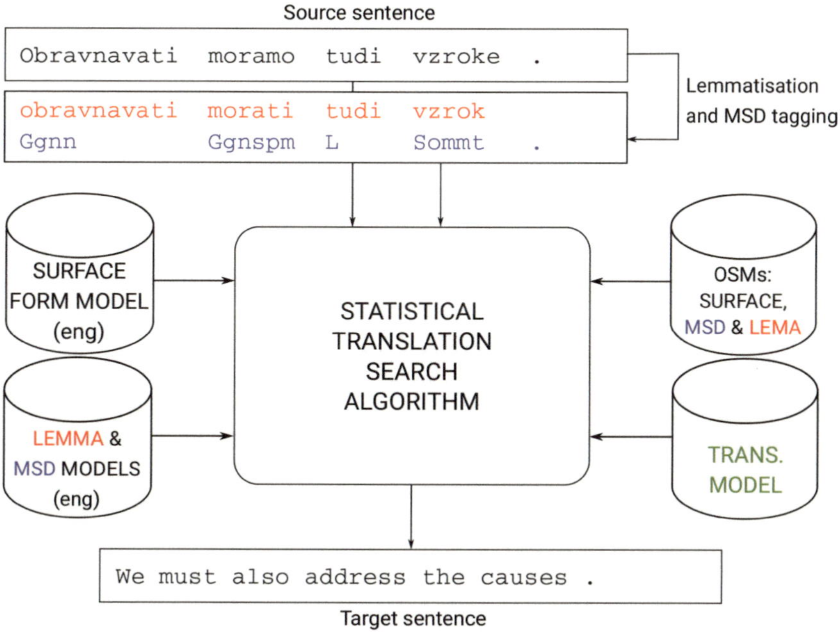

Figure 2.
Statistical machine translation system using a language model based on surface forms, a language model based on MSD tags, a language model based on lemmas, and three OSMs.

between the two approaches have narrowed, and hybrid approaches emerged, which try to gain benefit from both of them. We distinguish two groups of hybrid MT, those guided by rule-based MT and those guided by statistical approaches. Hybrid systems, guided by rule-based MT, use statistical MT to identify the set of appropriate translation candidates and/or to combine partial translations into the final sentence in the target language. Hybrid systems, guided by statistical MT, use rules at pre-/post-processing stages.

2.5 Neural machine translation

Neural MT emerged as a successor of statistical MT. It has made rapid progress in recent years, and it is paving its way into the translation industry as well. Neural MT is a deep learning-based approach to MT that uses a large neural network based on vector representations of words. If compared with statistical MT, there is no separate language model, translation model, or reordering model, but just a single sequence model, which predicts one word at a time. The prediction is conditioned on the source sentence and the already produced sequence in the target language. The prediction power of neural MT is more promising than that of statistical MT, as neural networks share statistical evidence between similar words. In **Figure 3** one of the proposed topology for neural machine translation is given with the same example sentence as in **Figure 2**. The input words are passed through the layers of the encoder (blue circles) to its last layer, the context vector, updating it for every input word. The context layer is then passed through the decoder layers (red circles) to output words, and it is again updated for each output word.

The encoder-decoder recurrent neural network architecture with attention is currently the state of the art for machine translation.

Although effective, the neural MT systems still suffer some issues, such as scaling to larger vocabularies of words and the slow speed of training the models.

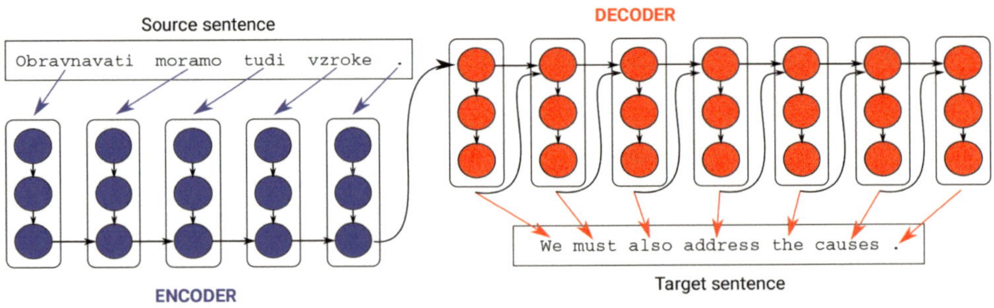

Figure 3.
Neural machine translation system using the encoder-decoder topology.

In addition, large corpus is needed to train neural MT systems with performance comparable to statistical machine translation. Researchers continue to work on solving open problems.

3. Problems in machine translation

The fast progress of MT has boosted translation quality significantly, but, unfortunately, machine translation approaches are not equally successful for all language pairs. Morphologically rich languages are problematic in MT, especially if the translation is from a morphologically less complex to a morphologically more complex language. Morphological distinctions not present in the source language need to be generated in the target language. Much work on morphology-aware approaches relies heavily on language-specific tools, which are not always available. Many morphologically rich languages fall in the category of low-resource languages.

One group of morphologically rich languages is a group of highly inflected languages. They are difficult not only for MT but also for other language technology applications [4, 5]. The main problem in highly inflected languages is that the large number of inflected word forms lead to data sparsity (see example in **Table 1**), which results in unreliable estimates in statistical MT [6]. Most words in a given corpus occur at most a handful of times. Therefore, the translation rule coverage is partial, and the estimation of translation probabilities is poor. Some approaches try to reduce the problem of data sparsity by using modelling units other than words; for example, stems and endings, lemmas and morphosyntactic tags, etc. Relaxed word order in inflectional languages poses another problem (see example in **Table 2**). Usually, very little information about the target word order is obtainable from the source sentence. Pre-ordering approaches learn to preprocess the source

	Singular	Dual	Plural
Nominative	študent	študenta	študenti
Genitive	študenta	študentov	študentov
Dative	študentu	študentoma	študentom
Accusative	študenta	študenta	študente
Locative	študentu	študentih	študentih
Instrumental	študentom	študentoma	študenti

In the example, the word has nine different endings.

Table 1.
Inflected word forms of the word "student" (masculine) in Slovene.

1.	Angleščino študiram dve leti.
2.	Dve leti študiram angleščino.
3.	Študiram angleščino, dve leti.
Third example is used in colloquial speech.	

Table 2.
Word permutations of the English sentence "I have been studying English for two years" in Slovene.

sentence during training in such a way that the words on the source side appear closer to their final positions on the target side. A frequent problem of inflectional languages is also an inaccurate translation of pronouns. There are also many cases in inflectional languages where the subject is dropped completely. Problematic also is differences in the expression of negation. Slavic languages fall into the category of highly inflected languages, and they cause many problems in machine translation [7, 8].

Another group of morphologically rich languages is a group of agglutinative languages, which are even more difficult to use in machine translation. In an agglutinative language, words may consist of more than one, and possibly many, morphemes. Each morpheme in a sequence indicates a particular grammatical meaning. Morphemes are used commonly as basic units in MT for those groups of languages. All these phenomena cause errors in translations produced by MT systems and make the use of MT questionable. It is necessary to evaluate MT quality before use in practice.

4. Machine translation evaluation

As MT emerges as an important mode of translation, its quality is becoming more and more important. Judging translation quality is called machine translation evaluation. It is defined commonly by technical terms. It means, with the exception of human (i.e. manual) evaluation, that it is defined as an algorithm that can be coded into a programme and run by a computer that calculates the evaluation score, which tells the user how good a translation is. Translation evaluation methods count word- and/or sentence-based errors that can be detected automatically, while general text-level aspects are not taken into account. This weakness of automatic MT evaluation is one of the main criticisms in the translation community. Despite that, in the last decade, we have been witnessing great progress in automatic MT evaluation.

MT quality can be measured in many different ways, depending on the goal of the evaluation and the means available. Traditionally, there are two paradigms of machine translation evaluation: Glass-box evaluation and black-box evaluation. Glass-box evaluation measures the quality of a system based on internal system properties. Black-box evaluation examines only the system output, without connecting it to the internal mechanisms of the translation system. The focus in this section will be on black-box evaluation. It is concerned only with the objective behaviour of the system upon a predetermined evaluation set. An evaluation set is a set of sentences in the source language and their translations into the target language, obtained by the translation system. These sentence pairs are then exposed to the evaluation. An evaluation set needs to be selected carefully to cover all data features important for future use of the translation system. The same translation quality can then be expected on the other data that is of the same type as the evaluation set; if not, translations of quite different quality could be obtained.

The reason is in the fact that MT systems are trained on translation examples. If these examples are of a different type to the text that is afterwards translated by the system, the system has only weak knowledge about its translation and, consequently, produces poor translations. Different types of data mean variations in structure, genre, and style. Evaluation, on the other hand, can focus on testing the systems's robustness. In this case, the evaluation set is composed of subsets of different data types. One should be aware that obtaining a robust MT system means at least training it with translation examples of different data types.

There is also a difference between judging and measuring the quality of MT output as a final product and judging and measuring the usability of MT output for subsequent corrections by humans, called post-editing (PE). As regards the latter, it is interesting to know how much editing effort is needed to make the MT output match a reference translation or to become an acceptable translation. What is acceptable translation is left to be decided by the translation expert. In the MT community, there are no criteria for it.

For a long time, methods for evaluating human and MT quality have been disconnected. The comparison between them was impossible. In recent years, a framework called multidimensional quality metrics (MQM) [9] was developed for evaluating the quality of both human and machine translations. It includes over 100 issue types that cover all of the major translation quality evaluation metrics. For the specific translation quality judgement task, relevant issues may be chosen from MQM. The focus of this section is only on the evaluation of MT quality, whereas human translation is taken as the gold standard.

4.1 Manual evaluation

The most common option for judging and measuring machine translation quality is human evaluation. The quality of MT output is judged by experts in translation and linguistics from two different perspectives. The first perspective is the degree of adherence to the target text and target language norms, referring, for example, to features such as grammaticality and clarity. This quality evaluation perspective is known as fluency. When judging fluency, the source text is not relevant. The evaluators have access to only the translation being judged and not the source data. Fluency requires an expert fluent only in the target language. On the other hand, source text adherence is judged to the source text norms and meaning, in terms of how well the target text represents the informational content of the source text. It is known as accuracy. The evaluators have access to the source text and translations being judged. Frequently, the context of a sentence is also taken into account. The evaluators must be bilingual in both the source and target languages. The adequacy and fluency are usually judged on a 5-point scale, as given in **Table 3**.

Human evaluation is time-consuming and expensive. It is also inherently subjective. To alleviate the problem of subjectivity, more experts are usually asked to evaluate the translations in the same evaluation set, and their evaluations are, finally, justified statistically.

4.2 Automatic evaluation

MT systems are rarely static, and they tend to be improved over time as resources grow and bugs are fixed. The evaluation needs to be repeated many times. Automatic evaluation metrics are cost-free alternatives to human evaluation. They are used commonly during the development of MT systems to estimate improvement. They are also applicable to compare different MT systems. While using automatic metrics to judge the translation quality, it is important to understand

Adequacy			Fluency
5	All meaning	5	Flawless language
4	Most meaning	4	Good language
3	Much meaning	3	Non-native language
2	Little meaning	2	Disfluent language
1	None	1	Incomprehensible

Table 3.
Numeric scale for judging adequacy and fluency.

what their scores mean. They rely on the idea that MT quality in itself should approach human quality. Automatic metrics depend on the availability of human reference translation. They evaluate the output of MT systems by comparing it to the reference translation. As there is a great variability even in human translation, it is important to have more human reference translations for each machine-translated sentence to be evaluated. Evaluation metrics then provide evaluation scores based on the most similar reference translation.

Standard and recently proposed automatic metrics for MT evaluation will be discussed in the continuation of this section. Statistical correlation coefficients are used to see how close automatic evaluation is to manual judgements. Three correlation coefficients will be described later in the section. Machine translation, coupled with subsequent post-editing, has become a widely accepted method in the translation industry. This type of translation workflow will be discussed at the end of this section.

5. Basic metrics for translation evaluation in MT

An obvious method for evaluation is to look at the translation and judge by hand, whether it is correct or not. To get reliable judgements, the evaluators should be appropriately qualified. From a practical point of view, manual evaluation, performed by translation experts, is expensive and takes time. What is needed are automatic metrics that are quick and cheap to use and approximate human judgements accurately. De facto metrics, used in the MT community, are BLEU, NIST, METEOR, and TER. All these metrics need reference translations because they compare the MT output with reference translations and provide comparison scores. If reference translations are available, these metrics can be used to evaluate the output of any number of systems quickly, without the need for human intervention. Let us take an example where the reference translation is "Dve leti že študiram angleščino" and the MT output to be evaluated is "Angleščino študiram dve leti". If we compute the precision, we get:

$$precision = \frac{correct}{length_o} = \frac{4}{4} = 100\%. \quad (3)$$

correct counts the number of correctly translated words, and $length_o$ is the length of machine translation output. For the same example, the recall is:

$$recall = \frac{correct}{length_r} = \frac{4}{5} = 80\%. \quad (4)$$

$length_r$ is the length of reference translation. F-measure results in:

$$F_1 = \frac{precision \cdot recall}{\frac{precision+recall}{2}} = 89\%. \quad (5)$$

Based on given measures, the quality of translation is good, as reordering is not penalised. It is not always a good decision. For example, the MT output "Dve angleščino študiram leti" will get the same evaluation result, even though the translation is disfluent.

BLEU [10] measures the overlap of unigrams (single words) and high-order n-grams between MT output and reference translations. It is defined as follows:

$$BLEU = \min\left(1, \frac{length_o}{length_r}\right)\left(\prod_{i=1}^{4} precision_i\right)^{\frac{1}{4}}. \quad (6)$$

The main component of BLEU is n-gram precision, i.e. $precision_i$. It is calculated as the ratio between matched n-grams and the total number of n-grams in the evaluated translation. Precision is calculated separately for each n-gram order, and the precisions are combined via a geometric averaging. The highest n-gram order is defined commonly to be four (four words in a sequence). Higher-order n-grams are used as an indirect measure of a translations level of grammatical well-formedness. The BLEU metric computes the modified precision score, weighted by the brevity penalty, which punishes sentences that are shorter than the reference. The final scores range from 0 to 1. **Table 4** contains the calculation of BLEU score for our example.

BLEU is typically computed over the entire corpus, not single sentences. It is important to point out that very few translations will attain a score of 1 unless they are identical to a reference translation. For this reason, even a human translator will not necessarily score 1, as there is great variability of possible correct translations. In this sense, it is also important to note that having more reference translations per sentence is highly welcome. It will increase the BLEU score. NIST [11] is a close derivate of BLEU.

Both metrics, BLEU and NIST, focus only on n-gram precision and disregard recall. Recall is the ratio between matched n-grams and the total number of n-grams in the reference translation. METEOR metric [12] combines precision and recall. The authors of METEOR argue that the brevity penalty in BLEU does not compensate adequately for the lack of recall. METEOR computes a score only for unigram

Metric	Score
1-gram precision	$\frac{4}{4}$
2-gram precision	$\frac{1}{3}$
3-gram precision	$\frac{0}{2}$
4-gram precision	$\frac{0}{1}$
Brevity penalty	$\frac{4}{5}$
BLEU	0%

Table 4.
BLEU score computation for the MT output "Angleščino študiram dve leti", if the reference is "Dve leti že študiram angleščino".

Reference:			dve	leti	že	študiram	angleščino
Output:	angleščino	študiram	dve	leti			
Edit:	I	I	M	M	D	D	D
$WER = \frac{0.5 \cdot 2 + 0.5 \cdot 3}{5} = 50\%$							

Table 5.
WER computation for the MT output "Angleščino študiram dve leti", if the reference is "Dve leti že študiram angleščino".

matching. Matching is done in three stages. The first stage is exact matching. Strings are aligned, which are identical in the reference and the translation. Words that are not matched are stemmed in the second stage. Stemming is the process of reducing inflected words to their word stem by cutting off the ends of words. Words with the same morphological root are aligned after stemming. In the last stage, unaligned words which are found to be synonyms are aligned, according to WordNet. WordNet [13] is a large lexical database of synonyms (called synsets). In WordNet, synsets are interlinked by means of conceptual-semantic and lexical relations. METEOR does not use higher-order n-grams, as n-gram counts do not require an explicit word-to-word matching. In METEOR, an explicit measure of the level of grammaticality is used. It captures directly how good the structure of the matched words in the machine translation is in relation to the reference.

Word error rate (WER) metric was first used to evaluate automatic speech recognition. It counts the minimum number of edits needed to change the evaluated translation so that it matches the references exactly, normalised by the average length of the references. The minimum number of edits is also called Levenshtein distance. Possible edits are insertion (I), deletion (D), and substitution (S) of single words. Matched words are denoted with M. Different edits can have different weights. For example, substitution is usually weighted at unity, but deletion and insertion are both weighted at 0.5:

$$WER = \frac{S + 0.5 \cdot D + 0.5 \cdot I}{length_r}. \tag{7}$$

$$TER = \frac{S + D + I + Shift}{length_r}. \tag{8}$$

Table 5 contains the calculation of WER for our example.

Translation edit rate (TER) metric [14] is a derivate from the WER. It uses an additional edit step, namely, shifts of word sequences (*Shift*). A shift moves a contiguous sequence of words within the evaluated translation to another location within the translation. All edits have equal cost. If more than one reference is available, and since the minimum number of edits needed to modify the translation is called for, only the number of edits to the closest reference is measured. TER is normalised by the average length of the reference. Position-independent error rate (PER) is another derivate from WER, which treats the reference and translation output as bags of words. Words from the translation are aligned to words in the reference, ignoring the position.

6. Advanced metrics for translation evaluation in MT

Although BLEU, NIST, METEOR, and TER metrics are used most frequently in the evaluation of MT quality, new metrics emerge almost every year. There is a metrics-shared task, held annually at the WMT Conference where new evaluation metrics are proposed [15, 16, 17]. Those which exhibit high correlation with human judgement will be presented from the pool of recently defined metrics.

CDER [18] is a more advanced metric that is concerned with edits and Levenshtein distance. It calculates the distance between two strings e_i^I and $e_i^{\sim L}$ using the auxiliary quantity $D(i,l)$, defined as:

$$D(i,l) := d_{CD}\left(e_1^i, e_1^{\sim l}\right). \tag{9}$$

$$D(0,0) = 0, \tag{10}$$

$$D(i,l) = min \left\{ \begin{array}{l} D(i-1,l-1) + c_{SUB}(e_i, \tilde{e}_l), \\ D(i-1,l) + 1, \\ D(i,l-1) + 1, \\ \min\limits_{i'} D(i',l) + 1. \end{array} \right\} \tag{11}$$

In addition to classical edit operations (i.e. insertion, deletion, and substitution), it models block reordering explicitly as an additional edit operation. As a further improvement, it introduces word dependent substitution costs $c_{SUB}(e_i, \tilde{e}_l)$. The observation that the substitution of a word with a similar one is likely to affect translation quality less than the substitution with a completely different word is accounted for in a metric score.

Tolerant BLEU [19] and LeBLEU [20] are derivates of BLEU with a relaxation of the strict word n-gram matching that is used in standard BLEU. Tolerant BLEU applies a specific distance measure that requires an exact match only in the middle of words, not in words as a whole. LeBLEU uses a distance measure based on characters. It also facilitates a fuzzy matching of longer chunks of text that allows, for example, to match two independent words with a compound.

CharacTER [21] is a derivate of TER:

$$CharacTER = \frac{S_{char} + D_{char} + I_{char} + Shift\ Cost}{length_{ochar}}. \tag{12}$$

It calculates the edit rate on character level, whereas shift edits are still performed on word level. First, a technique for word-level shifts is performed, words are then split into characters, and, finally, the edit distance is calculated based on characters, and *Shift Cost* is calculated. In addition, the lengths of translations in characters ($length_{ochar}$) instead of references are used for normalising the edit distance, which effectively counters the issue that shorter translations normally achieve lower TER. If we have two translations of different lengths, but with the same edit distance, they will obtain the same TER, as the length of the reference remains unchanged. In the same case, the longer translation will obtain lower TER if the edit distance is normalised by the length of translation.

METEOR universal [22] is a derivate of METEOR. It adds the fourth stage to matching. It is paraphrase matching. For each target phrase e_1, all source phrases f that e_1 translates are found. Each alternate phrase ($e_2 \neq e_1$) that translates f is considered a paraphrase with probability $P(f|e_1) \cdot P(e_2|f)$. The cumulative probability of e_2 being a paraphrase of e_1 is the sum over all possible pivot phrases f:

$$P(e_2|e_1) = \sum P(f|e_1) \cdot P(e_2|f). \tag{13}$$

Phrases are matched if they are listed as paraphrases in a language appropriate paraphrase table. Paraphrases are extracted automatically from the parallel corpora used to train statistical MT systems.

The BEER [23] metric provides a linear combination of different features:

$$BEER(h,r) = \sum_i w_i \times \phi_i(h,r) \tag{14}$$

where h is the system output and r is the reference. Each feature $\phi_i(h,r)$ has a weight w_i assigned to it. The first group of features consists of adequacy features. These features use precision, recall, and F1-score for different counts. F1-score is the harmonic mean of precision and recall multiplied by the constant of 2. The constant of 2 scales the F1-score to 1 when both recall and precision are 1. In BEER, function words and content words (nonfunction words) are counted separately. By differentiating function and nonfunction words, a better estimate is obtained of which words are more important and are less. The most important adequacy feature is a count of matching character n-grams. Using it, the translations are considered partially correct even if they did not get the morphology completely right. Character n-grams of order 6 are used. The second group of features comprises ordering features. Word order is evaluated by presenting the reordering as a permutation and calculating the distance to the ideal monotone permutation. Permutation trees are used to estimate long-distance reordering.

ChrF [24] is another, even simpler, metric based on character n-grams. It computes an F-score, based on precision and recall, using character n-grams:

$$ChrF\beta = \left(1+\beta^2\right)\frac{ChrP \cdot ChrR}{\beta^2 \cdot ChrP + ChrR} \tag{15}$$

ChrP and *ChrR* are character n-gram precision and recall, averaged over all n-grams. *ChrP* is the percentage of n-grams in the translation, which have a counterpart in the reference. *ChrR* is the percentage of character n-grams in the reference, which are also present in the translation. In the final score, the parameter is used, which gives more importance to recall than to precision. In [25], the optimal value for β is found to be 2, and the metric is called chrF2. In this metric, a recall has two times more importance than precision. WordF2 is a similar metric, where words are used instead of characters. Different weightings for n-grams were also investigated. Uniform weights are the most promising for machine translation evaluation.

DREEM [26] is a new metric based on distributed representations of words and sentences generated by deep neural networks. Neural networks are models that imitate human brains to recognise patterns in sequences. DREEM employs three different types of word and sentence representations: One-hot representations, distributed word representations learned from a neural network model, and distributed sentence representations computed with a recursive autoencoder. The final score is the cosine similarity of the representation of the translation and the reference, multiplied with a length penalty.

RATATOUILLE [27] is a metric combination of BLEU, BEER, METEOR, and few more metrics, out of which METEOR-WSD is a novel contribution. METEOR-WSD is an extension of METEOR that includes synonym mappings.

In this section, state-of-the-art MT evaluation metrics were investigated briefly. Only the most important characteristics of them were exposed. For a more elaborate description of each metric, the reader is advised to use the provided references to literature.

It should be noted that despite the well-known problems with BLEU, and the availability of many other metrics, MT system developers have continued to use BLEU as the primary measure of translation quality.

Today, different MT systems are available for use in practice. Usually, the qualities of different MT systems are compared between themselves by computing the translation quality scores on a predetermined evaluation set. The question arises whether, if there is a difference in quality on the evaluation set, one can be ensured that different MT systems indeed own different system quality. A difference in quality on an evaluation set may be just the result of happenstance. Research work on the statistical significance test for MT evaluation was done by Koehn [28], and the bootstrap resampling method is proposed to compute the statistical significance intervals for evaluation metrics on evaluation data. Statistical significance usually refers to the notions of the p-value, the probability that the observed difference in quality will occur by chance given the null hypothesis.

7. Correlation between automatic and human evaluation

Human judgements of translation quality are usually trusted as the gold standard, and the aim of an automatic evaluation metric is to produce quality estimates that are as close as possible to human judgements. As there are many different evaluation metrics, the user needs to decide which automatic evaluation metric he trusts the most. Correlation coefficients are used commonly to measure the closeness of automatic metric scores and manual judgements. Manual MT quality judgements on a number of test data are needed for comparison. Correlation coefficients are then computed on system level and/or segment level.

System-level comparison is done to compare different MT systems in general. First, each system gets a cumulative rank that reflects how high the annotators ranked that system. The metric scores of systems are also converted into ranks, and then the Spearman's rank correlation coefficient ρ is computed as [16]:

$$\rho = 1 - \frac{6 \cdot \sum_{i=1}^{n} d_i^2}{n \cdot (n^2 - 1)}. \tag{16}$$

d_i is the difference between the annotator's rank and metric's rank for system i. The number of systems is denoted with n. The possible values of ρ range between 1 (where all systems are ranked in identical order) and -1 (where the systems are ranked in the reverse order). Metrics with values of Spearman's ρ closer to 1 are better. The Spearman's correlation coefficient ρ is sometimes too harsh [17]: If a metric disagrees with humans in ranking two systems of a very similar quality, the ρ coefficient penalises this equally as if the systems were very distant in their quality. Pearson's correlation coefficient r is sometimes preferred [17]. It measures the strength of the linear relationship between a metric's scores and human scores:

$$r = \frac{\sum_{i=1}^{n} (H_i - \overline{H}) \cdot (M_i - \overline{M})}{\sqrt{\sum_{i=1}^{n} (H_i - \overline{H})^2} \sqrt{\sum_{i=1}^{n} (M_i - \overline{M})^2}}. \tag{17}$$

H is the vector of annotator's scores of all systems, and M is the vector of the corresponding scores as predicted by the given metric. $\overline{H}$ and $\overline{M}$ are their means, respectively. **Figure 4** shows Pearson's correlations of selected system-level metrics and MT systems built for different language pairs [15]. We can see that in majority of cases metrics correlate well with human judgements.

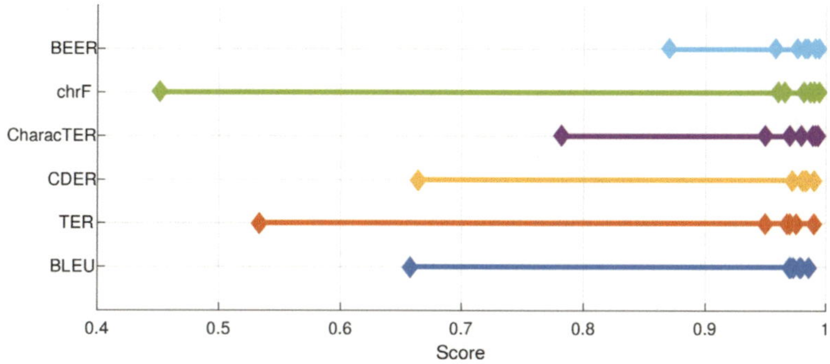

Figure 4.
Pearson's correlation coefficient r for selected evaluation metrics used for different MT systems.

The quality of a metric's segment level scores is usually measured by means of Kendall's τ rank correlation coefficient [17]. Let $r(\cdot)$ denotes annotator's rank and $m(\cdot)$ metric's rank. To compute Kendall's τ, the annotators rank all the translations of each segment from the best to the worst. Pairs (a, b) are then built where one system's translation $r(a)$ of a particular segment is judged to be (strictly) better than the other system's translation $r(b)$:

$$Pairs := \{(a,b) \,|\, r(a) < r(b)\}. \tag{18}$$

Afterwards, all concordant and discordant pairs are counted:

$$\begin{aligned} Con &:= \{(a,b) \in Pairs \,|\, m(a) > m(b)\}, \\ Dis &:= \{(a,b) \in Pairs \,|\, m(a) < m(b)\}. \end{aligned} \tag{19}$$

In a concordant pair, a human annotator and an automatic metric agree in ranking, and in a discordant pair, they disagree. Finally, Kendall's τ is computed as:

$$\tau = \frac{|Con| - |Dis|}{|Con| + |Dis|}. \tag{20}$$

τ value is between -1 (a metric always predicted a different order than humans did) and 1 (a metric always predicted the same order as humans). Metrics with higher τ are better.

In this section, no analysis of correlations for automatic metrics is presented, as it depends on many parameters. In general, all evaluation metrics presented in this chapter correlate well with human judgements. It is only worth mentioning that, for inflected languages, metrics that work on character level correlate better with human judgements than metrics that work only on word level.

8. Post-editing machine translation

In recent years, MT has become accepted more widely in the translation industry [29]. The most common workflow involves the use of machine-translated text as a raw translation that is corrected or post-edited by a translator. Post-editing (PE) tools and practices for such workflows are being developed in large multilingual organisations, such as the European Commission [30]. The researchers in [31] report that 30% of the companies in the translation industry currently use MT. The majority (70%) of the MT users combine MT with PE at least some of the time.

Post-editing MT is attractive because it has been shown to be faster than human translation. It is faster than translation from scratch and even faster than translation assisted by a translation memory [32]. Speed is not the only factor that should be taken into account when assessing the post-editing process. More recent studies have looked at ways of determining post-editing effort. In [33], three levels of post-editing effort are defined: Temporal effort, cognitive effort, and technical effort. The temporal effort is the time needed to post-edit a given text, cognitive effort is the activation of cognitive processes during post-editing, and the technical effort means the operations such as insertions and deletions that are performed during post-editing. All three levels of post-editing effort are influenced greatly by the translation quality. The use of PE and MT also raises the question about the quality of final translations. Has the quality improved, or is it worse?

As PE effort is related strongly to MT quality, derivatives of standard quality metrics are developed, which are concerned more with PE effort. Human-mediated translation error rate (HTER) [14] is a human-in-the-loop variant of TER. Instead of a reference, post-edited translation is used in the comparison. HTER centres on what edits are to be made to convert a translation into its post-edited version. It is computed as the ratio between the number of edit steps and the number of words in the post-edited version. HTER can be used as a measure of technical PE effort: The fewer changes necessary to convert the translation into its post-edited version, the less the effort required from the translator.

HTER is concerned more with the final translation and not the process. In [34] a metric called actual edit rate (AER) is proposed, which measures the translator's actual edit operations, which may involve more complex tasks, for example, applying corrections to previously post-edited parts of the text.

A study on PE of MT confirmed the relation between HTER and MT qualities [34]. An increase in HTER was evident as the quality of the MT system decreased. In contrast, they did not establish any significant association between AER and MT qualities. Keyboard activity may not be as sensitive to MT quality as PE time. They also found a linear relationship between MT quality and post-editing speed. MT quality was measured by the BLEU score of the system. The increase of BLEU score by one point resulted in a decrease of post-editing speed of about 0.16 seconds/word post-editing time. Their study also shows the correlation between the quality of machine translation output and the quality after post-editing. They confirmed that worse translation almost always leads to worse result after post-editing. As the use of MT and PE workflows has increased, there is a growing demand for expertise in PE skills. The research on and teaching of skills specific to post-editing has become necessary. The authors in [31] emphasise the impact of "familiarity with translation technology" on the employability of future translators.

9. Conclusion

Machine translation is being used by millions of people on a daily basis. This chapter discusses different MT approaches that were developed over time. Currently, the most promising approach is neural machine translation. Although effective, it also suffers some issues, such as scaling to larger vocabularies of words and the slow speed of training the models. Researchers continue to work on solving the problems and making translation a better service accessible to everyone.

The second part of the chapter describes how machine translation output is evaluated. The main characteristics of human and automatic MT evaluation were outlined. Human evaluation of MT output remains crucial to look for ideas to improve MT systems still further. On the other hand, automatic MT evaluation is cheap and fast. In the chapter, traditional and advanced metrics for automatic MT evaluation were presented. Despite the well-known problems with BLEU, and the availability of many other metrics, MT system developers have continued to use BLEU as the primary measure of translation quality.

MT quality is continually improving. Despite that, there are still a number of flaws in machine translation output. To make the translation correct, post-editing machine translation output is proposed to be integrated into the translation processes. It is discussed at the end of the chapter.

Future research in MT will be devoted to neural machine translation. It is still not very well understood. Its inner workings are commonly seen as a black-box, which works as the neurons of the human brain. As the computing power NMT requires becomes more widely available, many different configurations can be examined to further improve the accuracy of machine translation.

Future effort in machine translation evaluation will be directed toward character-based metrics which show the highest correlation with human judgement at the system and segment levels.

Human translators are worried to be replaced by machines. Machine translation, no matter how sophisticated, cannot match the accuracy of people. Human translators are also an important segment in MT evolution not only as post-editors but also as teachers for MT systems to become better and better.

Acknowledgements

The authors acknowledge the financial support from the Slovenian Research Agency (research core funding No. P2-0069).

Author details

Mirjam Sepesy Maučec* and Gregor Donaj
Faculty of Electrical Engineering and Computer Science, University of Maribor, Slovenia

*Address all correspondence to: mirjam.sepesy@um.si

References

[1] Doherty S, Gaspari F, Groves D, van Genabith J. Mapping the industry. I: Findings on translation technologies and quality assessment. In: GALA. 2013

[2] Koehn P, Och FJ, Marcu D. Statistical phrase-based translation. In: Proceedings of the 2003 Conference of the North American Chapter of the Association for Computational Linguistics on Human Language Technology, Vol. 1. Association for Computational Linguistics; 2003. pp. 48-54

[3] Durrani N, Schmid H, Fraser A. A joint sequence translation model with integrated reordering. In: Proceedings of the 49th Annual Meeting of the Association for Computational Linguistics: Human Language Technologies. Vol. 1. Association for Computational Linguistics; 2011. pp. 1045-1054

[4] Donaj G, Kačič Z. Language Modeling for Automatic Speech Recognition of Inflective Languages: An Applications-Oriented Approach Using Lexical Data. Springer; 2016

[5] Donaj G, Kačič Z. Context-dependent factored language models. EURASIP Journal on Audio, Speech, and Music Processing. 2017;**2017**(1):6

[6] Maučec MS, Donaj G. Morphosyntactic tags in statistical machine translation of highly inflectional language. In: Proceedings of the Artificial Intelligence and Natural Language Conference (AINL FRUCT); Saint-Petersburg, Russia. 2016. pp. 99-102

[7] Maučec MS, Brest J. Slavic languages in phrase-based statistical machine translation: A survey. Artificial Intelligence Review. 2019;**51**(1):77-117

[8] Maučec MS, Donaj G. Morphology in statistical machine translation from english to highly inflectional language. Information Technology and Control. 2018;**47**(1):63-74

[9] Lommel AR, Burchardt A, Uszkoreit H. Multidimensional quality metrics: A flexible system for assessing translation quality. In: Proceedings of ASLIB: Translating and the Computer. Vol. 35. 2013

[10] Papineni K, Roukos S, Ward T, Zhu W-J. Bleu: A method for automatic evaluation of machine translation. In: Proceedings of the 40th Annual Meeting on Association for Computational Linguistics. Association for Computational Linguistics; 2002. pp. 311-318

[11] Doddington G. Automatic evaluation of machine translation quality using n-gram co-occurrence statistics. In: Proceedings of the Second International Conference on Human Language Technology Research. Morgan Kaufmann Publishers Inc; 2002. pp. 138-145

[12] Lavie A, Agarwal A. Meteor: An automatic metric for MT evaluation with high levels of correlation with human judgments. In: Proceedings of the Second Workshop on Statistical Machine Translation. Association for Computational Linguistics; 2007. pp. 228-231

[13] Fellbaum C. WordNet. In: Poli R, Healy M, Kameas A, editors. Theory and Applications of Ontology: Computer Applications. Springer; 2010. pp. 231-243

[14] Snover M, Dorr B, Schwartz R, Micciulla L, Makhoul J. A study of translation edit rate with targeted human annotation. In: Proceedings of Association for Machine Translation in the Americas. Vol. 200. 2006

[15] Ma Q, Bojar O, Graham Y. Results of the wmt18 metrics shared task: Both

characters and embeddings achieve good performance. In: Proceedings of the Third Conference on Machine Translation: Shared Task Papers. 2018. pp. 671-688

[16] Macháček M, Bojar O. Results of the WMT13 metrics shared task. In: Proceedings of the Eighth Workshop on Statistical Machine Translation. 2013. pp. 45-51

[17] Macháček M, Bojar O. Results of the wmt14 metrics shared task. Proceedings of the Ninth Workshop on Statistical Machine Translation. 2014:293-301

[18] Leusch G, Ueffing N, Ney H. Cder: Efficient MT evaluation using block movements. In: 11th Conference of the European Chapter of the Association for Computational Linguistics. 2006

[19] Libovický J, Pecina P. Tolerant bleu: A submission to the wmt14 metrics task. In: Proceedings of the Ninth Workshop on Statistical Machine Translation. 2014. pp. 409-413

[20] Virpioja S, Grönroos S-A. Lebleu: N-gram-based translation evaluation score for morphologically complex languages. In: Proceedings of the Tenth Workshop on Statistical Machine Translation. 2015. pp. 411-416

[21] Wang W, Peter J-T, Rosendahl H, Ney H. Character: Translation edit rate on character level. In: Proceedings of the First Conference on Machine Translation: Volume 2, Shared Task Papers, Volume 2. 2016. pp. 505-510

[22] Denkowski M, Lavie A. Meteor universal: Language specific translation evaluation for any target language. In: Proceedings of the Ninth Workshop on Statistical Machine Translation. 2014. pp. 376-380

[23] Stanojevic M, Sima'an K. Beer: Better evaluation as ranking. In: Proceedings of the Ninth Workshop on Statistical Machine Translation. 2014. pp. 414-419

[24] Popović M. Chrf: Character n-gram f-score for automatic MT evaluation. Proceedings of the Tenth Workshop on Statistical Machine Translation. 2015: 392-395

[25] Popović M. Chrf deconstructed: Beta parameters and n-gram weights. In: Proceedings of the First Conference on Machine Translation: Volume 2, Shared Task Papers, Volume 2. 2016. pp. 499-504

[26] Chen B, Guo H, Kuhn R. Multi-level evaluation for machine translation. In: Proceedings of the Tenth Workshop on Statistical Machine Translation. 2015. pp. 361-365

[27] Marie B, Apidianaki M. Alignment-based sense selection in meteor and the ratatouille recipe. Proceedings of the Tenth Workshop on Statistical Machine Translation. 2015:385-391

[28] Koehn P. Statistical significance tests for machine translation evaluation. In: Proceedings of the 2004 Conference on Empirical Methods in Natural Language Processing. 2004

[29] Way A. Quality expectations of machine translation. In: Translation Quality Assessment. Springer; 2018. pp. 159-178

[30] Bonet J. No rage against the machine. Languages and Translation. 2013;**6**(2)

[31] Gaspari F, Almaghout H, Doherty S. A survey of machine translation competences: Insights for translation technology educators and practitioners. Perspectives. 2015;**23**(3):333-358

[32] Plitt M, Masselot F. A productivity test of statistical machine translation post-editing in a typical localisation context. The Prague Bulletin of Mathematical Linguistics. 2010;**93**:7-16

[33] Krings HP, Shreve GM. Repairing Texts: Empirical Investigations of Machine Translation Post-Editing Processes. Vol. 5. Kent State University Press; 2001

[34] Sanchez-Torron M, Koehn P. Machine translation quality and post-editor productivity. In: AMTA 2016 Vol. 2016. p. 16

Chapter 9

Pedestrian Detection and Tracking in Video Surveillance System: Issues, Comprehensive Review, and Challenges

Ujwalla Gawande, Kamal Hajari and Yogesh Golhar

Abstract

Pedestrian detection and monitoring in a surveillance system are critical for numerous utility areas which encompass unusual event detection, human gait, congestion or crowded vicinity evaluation, gender classification, fall detection in elderly humans, etc. Researchers' primary focus is to develop surveillance system that can work in a dynamic environment, but there are major issues and challenges involved in designing such systems. These challenges occur at three different levels of pedestrian detection, viz. video acquisition, human detection, and its tracking. The challenges in acquiring video are, viz. illumination variation, abrupt motion, complex background, shadows, object deformation, etc. Human detection and tracking challenges are varied poses, occlusion, crowd density area tracking, etc. These results in lower recognition rate. A brief summary of surveillance system along with comparisons of pedestrian detection and tracking technique in video surveillance is presented in this chapter. The publicly available pedestrian benchmark databases as well as the future research directions on pedestrian detection have also been discussed.

Keywords: pedestrian tracking, pedestrian detection, visual surveillance, pattern recognition, artificial intelligence

1. Introduction

The word surveillance, prefix sur is a French word means "over" and the root veiller means "to watch." In distinction to surveillance, Steve Mann in [1] introduces the term "sousveillance." Contrasting the word sur, sous meaning is "under," i.e., it signifies that the camera is with human physically (ex. camera mounting on head). Surveillance and sousveillance both are used for continuous attentive observation of a suspect, prisoner, person, group, or ongoing behavior and activity in order to collect information. In order to improve conventional security systems, the use of surveillance system has been increasingly emboldened by government and private organizations. Currently, surveillance systems have been widely investigated and used effectively in several applications like (a) transport systems (railway stations, airports, urban and ruler motorway road networks), (b) government agencies (military base camps, prisons, strategic infrastructures, radar centers, laboratories, and hospitals), (c) industrial environments, automated teller machine (ATM), banks, shopping malls, and public buildings, etc. The most of the surveillance systems at public and private places depend on the human operator observer, who detect any suspicious pedestrian activities in a video scene [2, 3]. The term pedestrian is a person who is walking or running on the street. In some communities, a person using wheelchair is also considered as pedestrians. The most challenging task for automatic video surveillance is to detect and track the suspicious pedestrian activity. For a real-time dynamic environment, the learning-based methods did not provide an appropriate solution for real-time scene analysis because it is difficult to obtain a prior knowledge about all the objects. Still, the learning-based methods are adopted due to their accuracy and robust nature. In the literature, several researchers use efficiently deep-learning (DL) based model for classification purpose in video surveillance over traditional approaches viz. perceptron model, probabilistic neural network (PNN), radial basis neural network (RBN), etc. Numerous learning-based techniques include artificial neural network (ANN), support vector machine (SVM), AdaBoost, etc. These techniques require the features such as histogram of oriented gradients (HOG), speeded-up robust features (SURF), local binary pattern (LBP), scale and invariant feature transform (SIFT), etc. to classify the type of object. Specifically, these features are represented by different deep learning algorithm versions such as the deep belief networks (DBN), recurrent neural network (RNN), generative adversarial networks (GANs), convolutional neural network (CNN), restricted Boltzmann machine (RBM), AlphaGo, AlphaZero, capsule networks bidirectional encoder representations for transformers (BERT), etc.

These variants of DL algorithms are used in many computer vision applications like face recognition, image classification, speech recognition, text-to-speech generation, handwriting transcription, machine translation, medical diagnosis, cars: drivable area, lane keeping, pedestrian and landmark detection for driver, digital assistants, ads, search, social recommendations, game playing, and content-based image retrieval. The advantage of DL approaches is its ability to learn complex scene features with very less processing of raw data and its capability of learning unlabeled raw data efficiently. Most recently, a new deep-learning technique called CNN have shown high performances over conventional methods in video processing research space. CNN can handle efficiently complex and large data.

During the past decade video surveillance systems have revolved from the simple video acquisition system to real-time intelligent autonomous systems. **Figure 1** shows a timeline chart of the evolution of video surveillance.

Visual surveillance systems come back into existence back in 1942. Primarily, closed-circuit television (CCTV) is used commercially as a security system, mainly for indoor environment. The main concerns of initial CCTVs were (1) voltage

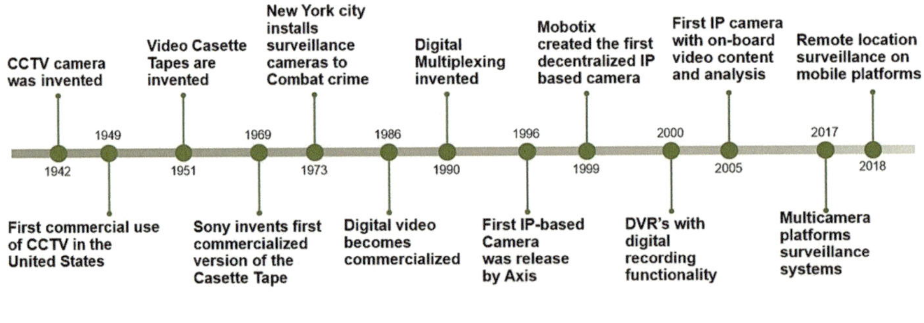

Figure 1.
Evolution of surveillance systems.

signals not openly transmitted in a distributed environment, (2) CCTV depends on strategic placements of cameras as per the geographical structure of workplace, (3) human observer is required for camera inputs to monitor the CCTV recorded footage [4]. The CCTV loses its primary advantage as an active, real-time medium, because the video footage can be used only after the fact or incident occurs, that can be used as a legal evidence or forensic tool. Next, in 1996, IP-based surveillance cameras were introduced by Axis, that overcomes the limitation of initial CCTV cameras such as (1) IP-based camera's transmits the raw images instead of voltage signals using the secure transmission channel of TCP/IP, (2) IP-camera comes along with the video analytics, i.e., camera itself can be used for analyzing the images, (3) Ethernet cable can be used as a medium for power supply instead of dedicated power supply, and (4) two-way bidirectional audio signals can be transmitted over a single dedicated network [5]. The recent surveillance system facilitates with remote location monitoring on handheld device like mobile phones.

The video surveillance systems can be categories based on a camera system, application and architecture. The camera system includes single camera, multi camera, fixed camera, moving camera and hybrid camera systems, etc. The application-based system includes object tracking and recognition, ID re-identification, customized event notification and alert based system, behavior analysis, etc. Finally, the architecture-based system includes standalone systems, cloud-based and distributed systems [6]. A general framework of automated visual surveillance system is shown in **Figure 2** [7–9]. Normally video surveillance system is based on multiple cameras, the videos from the multiple cameras are taken through the network and store in database.

The data need to be fused before incorporating the further processing. This can be done using data fusion techniques such as multi-sensory level, track to track and appearance to appearance [10–12]. After the data fusion following steps are performed. The traditional video surveillance system consists of various steps such

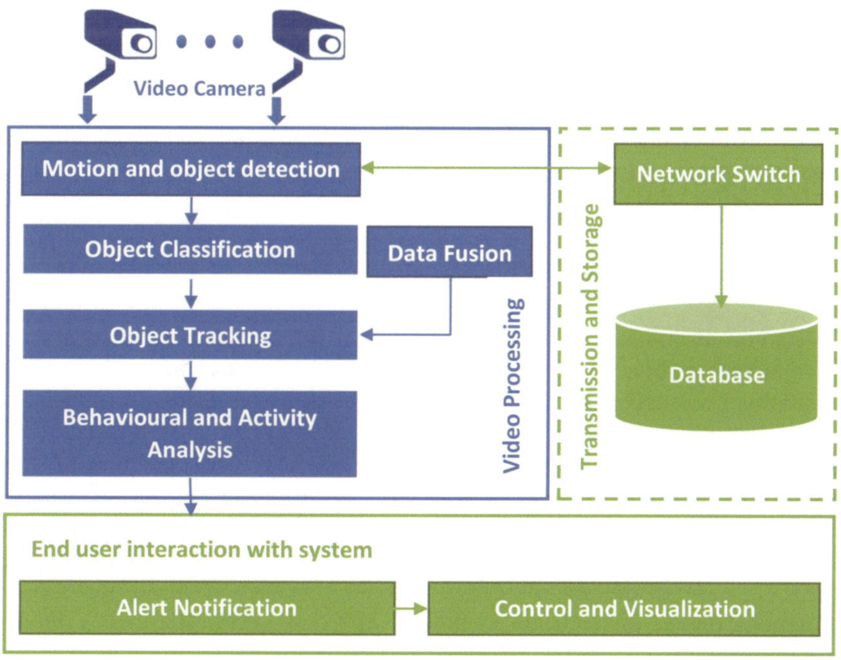

Figure 2.
A general framework of an automated visual surveillance system [7–9].

as (1) motion and object detection, (2) object classification, (3) object tracking, (4) behavior understanding and activity analysis, (5) pedestrian identification and (6) data fusion. Each stage of automated visual surveillance system is described as follows.

1.1 Motion and object detection

Object detection is the first step that deals with detecting instances of semantic objects of a certain class, such as humans, buildings, cars, etc. in a sequence of videos. The different approaches of object detection are frame-to-frame difference, background subtraction and motion analysis using optical flow techniques [13]. These approaches typically use extracted features and learning algorithms to recognize instances of an object category. The object detection process is divided into two categories. First, object detection, which include mainly three types of methods such as background subtraction, optical flow and spatiotemporal filtering. Second, object classification, use primarily visual features as shape based, motion based and texture-based method [13]. Motion detection is one of the problems in video surveillance, as it is not only responsible for the extraction of moving objects, but also critical to many applications including object-based video encoding, human motion analysis, and human machine interactions [14, 15].

After object detection, next step is motion segmentation. This step is used for detecting regions corresponding to moving objects such as humans or vehicles. It mainly focuses on detecting moving regions from video frames, and creating a database for tracking and behavior analysis. Motion detection is used for detecting a change in the position of an object, relative to its surroundings or a change in the surroundings, relative to an object. Motion detection can be achieved using electronic motion sensors, which detect the motion from the real environment.

1.2 Object tracking

Tracking of objects in a video sequence means identifying the same object in a sequence of frames using the object unique characteristics represented in the form of features. Generally, the detection process is always followed by tracking in video surveillance systems. Tracking is performed from one frame to another, using tracking algorithms such as kernel-based tracking, point based tracking and silhouette-based tracking [16].

1.3 Behavior and activity analysis

In some conditions, it is mandatory to analyze the behaviors of people and determine whether their behaviors are suspicious or not, such as the behavior of pedestrian at a crowded place (e.g. public market places and government offices, etc.). In this step the motion of objects is recognized from the video scene and generate the description of the action. Ahmed Elaiw et al. [80] proposed a critical analysis and modelling strategy of human crowds with the intention of selecting the most relevant scale out of three approaches, i.e., (1) microscopic, means pedestrian are individual detected based on the location, velocity and motion parameter is neglected, (2) mesoscopic, means pedestrian are detected based on position, velocity and depend on the distribution function and (3) macroscopic, mean the pedestrian are identified based on the average pedestrian quantity, moment of pedestrian. It can be used for efficient decision making in critical situations when human crowd safety is important. Safety of human crowds depends upon the quantity and density of pedestrian move physically at different high crowed places.

1.4 Person identification

The last step is human identification. Human face and gait are the main biometric features that can be used for personal identification in visual surveillance systems after a behavior analysis [8].

The goal of this chapter is to discuss the issues and challenges involved in designing visual surveillance system. Again, group pedestrian detection and tracking methods used for moving and fixed camera into broad categories and give an informative analysis of relative methods in each category. The main contributions of this chapter are as follows:

- The comparative analysis of publicly available benchmark datasets of pedestrian with its use, specification and environment limitation

- Analyze issues and challenges of pedestrian detection and tracking in the video sequences captured by a moving and fixed camera

- Categorizing the methods of pedestrian detection and tracking in different ways based on the general concept of methods belonging to each category and described proposed improvements for each method

This chapter is organized into the following sections. Section 1 gives an introductory part, the importance of video surveillance system, recent advancement and general framework of video surveillance. Section 2, discusses different benchmark pedestrian datasets used to compare the different methods of pedestrian detection and tracking. Section 3, presents a detailed discussion on issues and challenges of pedestrian detection and tracking in video sequence. Section 4, groups the methods of pedestrian detection and tracking method for moving and fixed camera into different categories, describe their general concept with the improvements in each category. In Section 5, discusses possible future directions. Finally, the chapter concluded with a discussion in Section 6.

2. Pedestrian datasets reported in literature

The state-of-the-art methods for pedestrian detection and tracking method include adaptive local binary pattern (LBP), histogram of oriented gradient (HOG) into a multiple kernel tracker, spatiotemporal context information-based method using benchmark databases [10]. In this section we outlined the benchmark datasets that has been commonly used by the researchers. **Figure 3** shows a sample image of each pedestrian dataset. Next, we discuss each database with its specification, use and environmental constrain followed by comparative analysis.

2.1 Massachusetts Institute of Technology (MIT) pedestrian dataset

It is one of the first pedestrian datasets, fairly small and relatively well solved at this point. This data set contains 709 pedestrian images taken in city streets. Out of this 509 training and 200 test images of pedestrian in city scenes. Each image contains either a front or a back view with a relatively limited range of poses [11, 12].

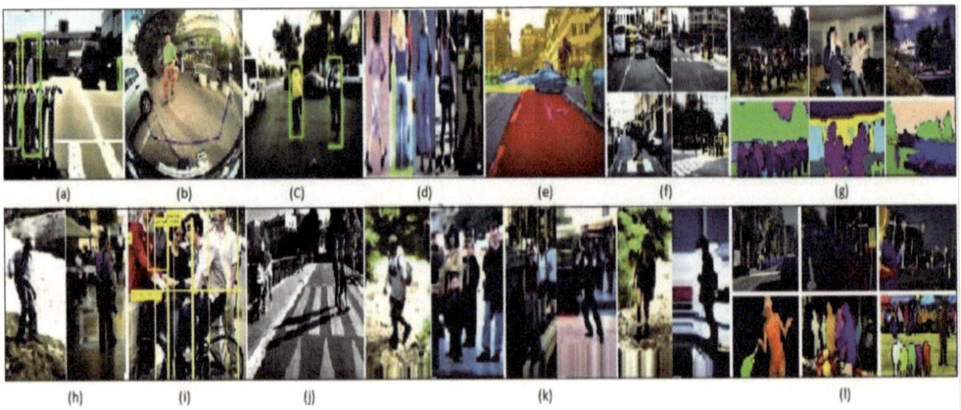

Figure 3.
Example of pedestrians dataset. (a) Caltech pedestrian dataset images consists of unique annotated pedestrians. (b) GM-ATCI rear-view pedestrians' dataset. (c) Tsinghua-Daimler Cyclist Detection Benchmark dataset images. (d) NICTA urban dataset. (e) ETH urban dataset. (f) TUD-Brussels dataset. (g) Microsoft COCO pedestrian dataset. (h) INRIA person static pedestrian detection datasets. (i) PASCAL object dataset. (j) CVC-ADAS collection of pedestrian datasets. (k) MIT pedestrian database images. (l) Mapillary vistas research dataset.

2.2 Caltech pedestrian dataset

The Caltech dataset consists of 640 × 480 resolution video taken from a vehicle driving through regular traffic in an urban environment. About 250,000 with a total of 350,000 bounding boxes and 2300 unique pedestrians were annotated for testing and training purpose. The annotation includes bounding boxes for each pedestrian walking on streets and detailed occlusion labels for each object captured in a video sequence in an urban environment. The annotation of pedestrians is used for validating the pedestrian detection and tracking algorithm accuracy [10].

2.3 General Motors-Advanced Technical Center (GM-ATCI) pedestrian dataset

GM-ATCI dataset is a rear-view pedestrians database captured using a vehicle-mounted standard automotive rear-view display camera for evaluating rear-view pedestrian detection. In total, the dataset contains 250 clips duration of 76 min and over 200K annotated pedestrian bounding boxes. The dataset has been captured at different locations, including: indoor and outdoor parking lots, city roads and private driveways. This dataset was collected in both day and night scenarios, with different weather and lighting conditions [15].

2.4 Daimler pedestrian dataset

The pedestrian images captured from a vehicle-mounted calibrated stereo camera rig in an urban environment. This dataset contains tracking information and a large number of labeled bounding box with a float disparity map and a ground truth shape image. The training set contains 15,560 pedestrian samples with 6744 label pedestrian and testing set contains more than 21,790 images with 56,492 pedestrian labels [15].

2.5 National Information and Communication Technology Australia (NICTA) pedestrian dataset

It is a large-scale urban dataset collected in multiple cities and countries. The dataset contains around 25,551 unique pedestrians of humans, allowing for a dataset of over 50 K images with mirroring and annotation for validating detection and tracking algorithm accuracy [16].

2.6 Swiss Federal Institute of Technology (ETH) pedestrian dataset

It is an urban dataset captured from a stereo rig mounted on a stroller. Observing a traffic scene from inside a vehicle. The database is used for pedestrian detection and tracking from moving platforms in an urban scenario. Dataset consists of traffic agents such as different cars and pedestrians. One can predict their further motion, or even interpret their intention. At the same time, one needs to stay clear of any obstacles, remain on the assigned road, and read or interpret any traffic signs on the side of the street. On top that, a human is able to assess the situation, when close to a school or pedestrian crossing, one ideally will adapt one's driving behavior [17].

2.7 TUD-Brussels pedestrian dataset

This dataset consists of pairs recorded in a crowded urban setting from a moving platform with an onboard camera and challenging automotive safety scenario in urban environment [18].

2.8 National Institute for Research in Computer Science and Automation (INRIA) pedestrian dataset

INRIA is currently one of the most popular static pedestrian detection datasets. It contains moving people with significant variation in appearance, pose, clothing, background, illumination, coupled with moving cameras and backgrounds. Each pair shows two consecutive frames [19].

2.9 PASCAL visual object classes (VOC) 2007 and 2012 dataset

This is static object dataset with diverse object views and poses. The goal of visual object classes challenge is to recognize objects from a number of visual object classes in realistic scenes. The 20 object classes that have been selected are (1) person, (2) animal, (3) vehicle [20].

2.10 Microsoft Common Object in Context (COCO) 2018 dataset

The COCO is recent dataset created by Microsoft [22]. The dataset designed to spur object detection research with a focus on detecting objects in context. The annotations include different instances of segmentations for objects belonging to 80 categories of object, stuff segmentations for 91 categories, key point annotations for person instances, and five image label per image. The different COCO 2018 dataset challenges are (1) object detection with segmentation masks on the image, (2) panoptic segmentation, (3) person key point estimation, and (4) dense pose detection. **Figure 3(g)** shows the sample images of MS COCO dataset.

2.11 Mapillary vistas research dataset

The Mapillary vistas panoptic segmentation targets the full perception stack for scene segmentation in street-images [22]. Panoptic segmentation solves both stuff and thing classes, unifying the typically distinct semantic and instance segmentation tasks efficiently. **Figure 3(l)** shows a sample image of Mapillary vistas research datasets. The comparative analysis of recently utilized pedestrian database with its

Data source	Purpose	Image or video clips	Annotation	Environment	Ref.	Year
MIT	City street pedestrian segmentation, detection and tracking	709 pedestrian images 509 training and 200 test images	No annotated pedestrian	Day light scenario	[11] [12]	2000, 2005
Caltech pedestrian dataset	Detection and tracking of pedestrian walking on the street	250,000 frames (in 137 approximately minute long segments)	350,000 bounding boxes and 2300 unique pedestrians were annotated	Urban environment	[10]	2012
GM-ATCI	Rear view pedestrian segmentation, detection and tracking	250 video sequences	200K annotated pedestrian bounding boxes	Dataset was collected in both day and night scenarios, with different weather and lighting conditions	[13]	2015
Daimler	Detection and tracking of pedestrian	15,560 pedestrian samples, 6744 negative samples	2D bounding box overlap criterion and float disparity map and a ground truth shape image	Urban environment	[15]	2016
NICTA 2016	Segmentation, pose estimation, learning of pedestrian	25,551 unique pedestrians, 50,000 images	2D ground truth image	Urban environment	[16]	2016
MS COCO 2018	Object detection, segmentation, keypoint detection, DensePose detection	300,000, 2 million instances, 80 object categories	5 captions per image	Urban environment	[22]	2018
Mapillary vistas dataset 2017	Semantic understanding street scenes	25,000 images, 152 object categories	Pixel-accurate and instance-specific human annotations for understanding street scenes	Urban environment	[22]	2017
MS COCO 2017	Recognition, segmentation, captioning	328,124 images, 1.5 million object instances	Segmented people and objects	Urban environment	[22]	2017
MS COCO 2015	Recognition, segmentation, captioning	328,124 images, 80 object categories	Segmented people and objects	Urban environment	[22]	2015
ETH	Segmentation, detection, tracking	Videos	Dataset consist of other traffic agents such as different cars and pedestrians	Urban environment	[17]	2010
TUD-Brussels	Detection, tracking	1092 image pairs	1776 annotated pedestrian	Urban environment	[18]	2009

Data source	Purpose	Image or video clips	Annotation	Environment	Ref.	Year
INRIA	Detection, segmentation	498 images	Annotations are marked manually	Urban environment	[19]	2005
CVC-ADAS	Detection, tracking	60,000 frames	7900 annotated pedestrians	Urban environment	[20]	2009
PASCAL VOC 2012	Detection, classification, segmentation	11,530 images, 20 object classes	27,450 ROI annotated 6929 segmentations	Urban environment	[21]	2012

Table 1.
Recently used pedestrian databases by the researchers.

application for video surveillance system is shown in **Table 1**. The comparison is performed in terms of application of dataset, size of dataset, dataset creation environment scenarios and type of annotation details used for testing, training and validation of detection and tracking algorithm performance. These datasets used by the researchers for testing the performance of their respective pedestrian detection and tracking algorithm.

3. Issues and challenges of pedestrian detection and tracking

The moving object is a nonrigid thing that moves over time in image sequences of a video captured by a fix or moving the camera. In video surveillance system the region of interest is a human being that needs to be detected and tracked in the video [23]. However, this is not an easy task to do due to the many challenges and difficulties involved. These challenges occur at three different levels of pedestrian detection. Video acquisition, human detection and its tracking. The challenges in acquiring video are, viz. illumination variation, abrupt motion, complex background, shadows, object deformation, etc. Human detection and tracking challenges are varied poses, occlusion, crowd density area tracking, etc. Each issues and challenges are represented here in this section.

3.1 Problems related to camera

Many factors related to video acquisition systems, acquisition methods, compression techniques, stability of cameras (or sensors) can directly affect the quality of a video sequence. In some cases, the device used for video acquisition might cause limitation for designing object detection and tracking (e.g., when color information is unavailable, or when the frame rate is very low). Moreover, block artifacts (as a result of compression) and blur (as a result of camera's vibrations) reduce the quality of video sequences [36]. Noise is another factor that can severely deteriorate the quality of image sequences. Besides, different cameras have different sensors, lenses, resolutions and frame rates producing different image qualities. A low-quality image sequence can affect moving object detection algorithms. **Figure 4** shows an example of each challenge.

3.1.1 Camera motion

When dealing with detecting moving objects in the presence of moving cameras, the need for estimating and compensating the camera motion is evitable. However,

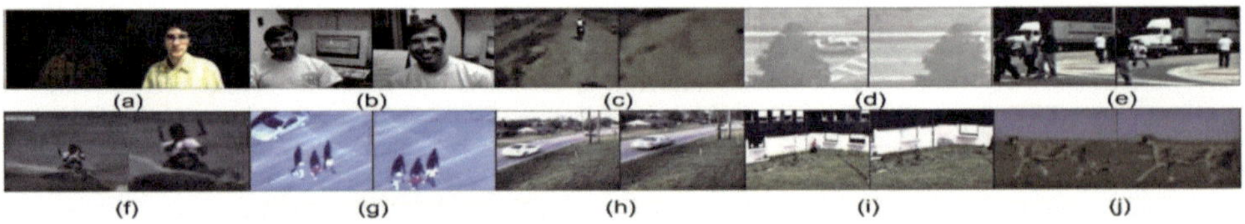

Figure 4.
Issues and challenges. (a) An example of illumination variation challenge (David indoor in the Ross dataset [39]). (b) An example of appearance change challenge (Dudek in the Ross dataset [39]). (c) An example of abrupt motion challenge (Motocross in the Kalal dataset [50]). (d) An example of occlusion challenge (car in the Kalal dataset [50]). (e) An example of freely motion of camera in the Michigan University dataset [10]. (f) An example of dynamic background challenge (Kitesurf in the Zhang dataset [60]). (g) An example of shadow challenge (pedestrian 4 in the Kalal dataset [50]). (h) An example of panning in camera in the CDNET database [10]. (i) An example of zooming in camera in the CDNET database [10]. (j) An example of nonrigid moving object in a video sequence [67].

it is not an easy task to do because of possible camera's depth changes and its complex movements. Many works elaborated an easy scenario by considering simple movements of the camera, i.e., pan tilt zoom (PTZ) cameras. This limited movement allows using a planar homography in order to compensate camera motions, which results in creating a mosaic (or a panorama) background for whole frames of the video sequence [37].

3.1.2 Nonrigid object deformation

In some cases, different parts of a moving object might have different movements in terms of speed and orientation. For instance, a walking dog when wags its tail or a moving tank when rotates its turret. When dealing with detecting such moving objects, most algorithms, different moving objects. It produces an enormous challenge, especially for nonrigid objects and in the presence of moving cameras. In Hou et al. [40], articular models have been proposed for moving nonrigid objects to handle nonrigid object deformation. In these models, each part of an articulated object is allowed to have different movements. It can be concluded that local features of a moving object along with updating background models are more efficient for dealing with this challenge.

3.2 Challenges in video acquisition

3.2.1 Illumination variation

The lighting conditions of the scene and the target might change due to the motion of light source, different times of day, reflection from bright surfaces, whether in-outdoor scenes, partial or complete blockage of the light source by other objects, etc. The direct impact of these variable results in background appearance changes, which causes false positive detections for the methods based on background modeling. Thus, it is essential for these methods to adapt their model to this illumination variation. Meanwhile, because the object's appearance changes under illumination variation, appearance-based tracking methods may not be able to track the object in the sequence [23–28]. Thus, it is required for these methods to use features which are invariant to illumination.

3.2.2 Presence of abrupt motion

Sudden changes in the speed and direction of the object's motion or sudden camera motion are another challenge of video acquisition that affects the object detection and tracking. If the object or the camera moves very slowly, the temporal differencing methods may fail to detect the portions of the object coherent to background [31]. Meanwhile, a very fast motion produces a trail of the ghost detected region. So, if this object's motions or camera motions are not considered, the object cannot correctly be detected correctly by methods based on background modeling. On the other hand, for tracking-based methods, prediction of motion becomes hard or even impossible; as a result, the tracker might lose the target. Even if the tracker does not lose the target, the unpredictable motion can introduce a greater amount of error in algorithms [32].

3.2.3 Complex background

The background may be highly textured, especially in natural outdoor environments where high variability of textures is presented in outdoor scenes. Moreover, the background may be dynamic, like it may contain movement (e.g., a fountain, clouds in movement, traffic lights, trees waggle, water waves, etc.). These need to be considered as background in many moving object detection algorithms. Such movements can be periodic or nonperiodic [34].

3.2.4 Shadows

The presence of shadows in video image sequences complicates the task of moving object detection. Shadows are created due to the occlusion of the object by the light source. If the object does not move during the sequence, resulted shadow is considered as static and can effectively be incorporated into the background. However, a dynamic shadow, caused by a moving object, has a critical impact for accurately detecting moving objects, since it has the same motion properties as the moving object and is tightly connected to it. Shadows can be often removed from images of the sequence using their observed properties such as color, edges and texture or applying a model based on prior information such as illumination conditions and moving object shape [35, 47, 48]. However, dynamic shadows are still difficult to be distinguished from moving objects, especially for outdoor environment where the background is usually complex.

Next, human detection and tracking issues and challenges are discussed in brief. It includes varying poses, occlusion, crowd density area tracking, etc.

3.3 Challenges in human detection and tracking

3.3.1 Pedestrian occlusion

The object may be occluded by other objects in the scene. In this case, some parts of the object can be camouflaged or just hidden behind other objects (partial occlusion) or the object can be completely hidden by others (complete occlusion). As an example, consider the target to be a pedestrian walking on the sidewalk. It may be occluded by trees, cars in the street, other pedestrians, etc. Occlusion severely affects the detection of objects in background modeling methods, where the object is completely missing or separated into unconnected regions [33]. If occlusion occurs, the object's appearance model can change for a short time, which can cause some of the object tracking methods.

3.3.2 Pose variation: moving object appearance changes

In real scenarios, most objects can occur in 3D space, but we have the projection of their 3D movement in a 2D plane. Hence, any rotation in the direction of third

Challenge	Proposed methodology	Advantage	Identifies gap	Observation	Ref.	Year
Problems related to camera	1. Heterogeneous feature-based technique 2. Color-based appearance model and the APSO based framework 3. Pyramidal structure model 4. Blur-driven tracker (BLUT) framework 5. Cascade particle filter in tracking	1. Detecting human movements in low resolution conditions 2. Deal with low resolution video sequences, the techniques based on fusion can be used to better detect moving objects 3. Fast detect moving objects 4. Can robustly track severely blurred targets 5. Detection for low frame rate videos	1. Low-resolution humans' images may not cover all targets with different sizes 2. Human detection accuracy is less in complex background 3. Not tested for real time scenarios 4. Misclassification in detection and tracking in complex background 5. Tracker not able to distinguish between different targets of video sequence	1. Region of Interest method can be used in outdoor environment 2. Appearance model, the object is significantly fast 3. Detect the moving object with detection accuracy are 80% 4. Effectively tracks the blurred objects without deblurring 5. Efficient multi-target tracker is required	[53] [52] [51] [50] [49]	2017 2015 2014 2011 2008
Camera motion	1. Image stabilization techniques based on motion compensation of features 2. 3D motion models 3. Multi plane representation of the 3D scene 4. Adaptive motion model	1. Efficiently detect moving object and motion blur issue resolve 2. Can be efficiently used to compensate camera vibrations 3. Computational complexity is decreased 4. More accurately detects moving objects	1. Motion detection fails in long video. It computationally complex 2. It cannot be applied to real-time systems due to the slowness of the SIFT computation 3. It is computationally very complex 4. Camera motion detection time is more	1. Adaptive particle filter framework uses PCA to reduce SIFT feature 2. 3D camera motion model requires fast feature extraction 3. It can be used for image analysis 4. Efficiently resolve the issue of motion blur	[63] [62] [61] [60] [59] [58] [57] [56] [55] [54]	2014 2011 2009 2009 2017 2005 1994 2017 2000 1998
Nonrigid object deformation	1. Target object as a combination of different segments having different movements 2. Model of articulated objects composed of rigid bodies 3. View-based eigenspace representation	1. Segments with motion consistency 2. Accurately produce a depth image representing different poses of a moving object 3. Produced good results for handling tracking	1. Computationally intensive 2. Dense articulated real-time tracker requires initial object pose. Tracker not able to measure the similarity of the object accurately for small object	1. Eigen tracker is used to track and recognize gestures 2. Accuracy of tracker can be improved by more color features 3. Tracker not able to detect the small objects	[66] [65] [64]	2017 2015 1998

Challenge	Proposed methodology	Advantage	Identifies gap	Observation	Ref.	Year
Illumination variation	1. Conditional background update scheme 2. Local representation model 3. 2D-Cepstrum approach 4. Adaptive local binary pattern 5. Bayesian framework 6. Color modelling approach with B-spline curves	1. Evaluate rapidly scene changes adaptively 2. More efficient to detect moving objects 3. Provided good robustness to illumination variations 4. Tolerant to illumination variations 5. Not sensitive to illumination variation 6. Adapt to irregular illumination variations and abrupt changes of brightness	1. Computation time required for this model is more 2. Final segmentation results consist of large noise patches 3. This method is computationally intensive 4. This method requires a nonmoving camera 5. If the colors changes are very fast then, this method fails to detect object 6. Speed of visual tracker is less. Visual tracker can detect only single target object	1. Results: TPF—7.36 ms and Speed FPS—136 2. Post processing can reduce large noise patches 3. Color and texture information retain 4. Texture features having FPS—15 fps 5. Dynamic texture coefficient is used for texture variations with 5 fps 6. Visual tracker speed improved using good workstation	[23] [24] [25] [26] [27] [28]	2017 2015 2010 2006 2006 2001
Presence of abrupt motion	1. Kernelized correlation filter tracker based on swarm intelligence method 2. Hamiltonian Markov Chain Monte Carlo (MCMC) based tracking algorithm 3. Bayesian filter tracking frame framework 4. Wang-Landau Monte Carlo sampling method	1. Effectively handle abrupt motion tracking in videos 2. Effective in handling different type of abrupt motions 3. Effective against abrupt motions 4. Efficiently deals with the abrupt motions	1. Kernelized correlation filter tracker computational complexity is more 2. Tracker does not handle abrupt motion in scale and position 3. Tracker not suitable in significant motion 4. Tracker does not fully consider the abrupt changes	1. A unified framework track smooth or abrupt motion 2. Tracker can handle smooth and abrupt motion 3. Bayesian filtering solve the local-trap problem 4. Tracking algorithm efficiently handle abrupt motions	[31] [30] [29] [28]	2017 2012 2012 2008
Complex background	1. Adaptive background model 2. Dynamic texture modelling methods 3. Principal features based on statistical characteristics 4. Auto-regressive model	1. Highly complex backgrounds for detecting moving objects 2. Effectively used for moving object detection 3. Effectively overcome the complex background	1. Less accurate in background regions with large re-projection errors 2. It is computationally very complex 3. Wrongly absorb a foreground object with background 4. The model is very complex	1. Adaptive to any object detection methods 2. Operates on dynamic texture sequence 3. Fusion of information solved the issue	[42] [41] [40] [39] [38] [37]	2017 2017 2005 2004 2003 1995

Challenge	Proposed methodology	Advantage	Identifies gap	Observation	Ref.	Year
	5. Active contour model	4. Effectively handle dynamic background 5. Can detect active segments	5. Output with noise patches	4. Filters operates with linear prediction model 5. It detects complex background object with noise patches		
Shadow	1. Shadow elimination algorithm using HSV color space and texture features 2. Modified Gaussian mixture model	1. Can effectively distinguish shadow regions 2. Can handle a highly dynamic environment object detection for moving camera	1. Long video accuracy reduces due to texture variation 2. Misclassification in shadow detection result if complex scene	1. Discrimination in image get difficult due to texture variation 2. This algorithm works in real time with good accuracies	[44] [43]	2017 2016
Pedestrian occlusion	1. Histogram of oriented gradient (HOG) into a multiple kernel tracker 2. Spatiotemporal context information-based method 3. Maintaining appearance models 4. Tracking was achieved by evolving the contour from frame to frame 5. Appearance model based on the filter responses from a steerable pyramid	1. Effectively handle occlusions in different conditions of moving cameras 2. Tracker is able to distinguish the object in occlusions effectively 3. Occlusions can be more efficiently handled 4. Minimize some energy function 5. Overcome changing appearance and occlusion problems	1. Histogram of oriented gradient (HOG) into a multiple kernel tracker 2. Spatiotemporal context information method is complex 3. Occlusion fail for long sequences with varying lighting conditions 4. Offline object tracking not possible for all types of objects 5. Tracked object move with its background. Tracker, fail when object is occluded	1. Effectively handle occlusions in moving cameras 2. Spatiotemporal context to analyze occlusion 3. Tracker deals with complex real scenario 4. Nonrigid object tracking derived by Bayesian framework 5. Motion-based tracker with RMS error rate is 0.1–1.1	[36] [35] [34] [33] [32]	2017 2007 2006 2004 2003
Pose variation—moving object appearance changes	1. Trainable model which uses the optical flow 2. Wandering-stable-lost framework model	1. Good performance in handling appearance changes 2. Adaptive and significant to appearance changes	1. Difficult to identify ambiguous motion pattern of object 2. Sensitive to lighting changes. Multiple cameras required to cope with self-occlusion	1. CNN, learn with synthetic video with Mean—69.7	[27] [26] [25] [24]	2017 2006 2006 2004

Challenge	Proposed methodology	Advantage	Identifies gap	Observation	Ref.	Year
	3. Covariance matrix and Lie algebra 4. Low-dimensional subspace representation	3. Adaptively track moving objects under their appearance changes for moving camera 4. Efficiently adapts online to changes in the appearance of the moving objects	3. Covariance tracker is computationally intensive 4. Tracker occasionally drift from a target object	2. Models appearance using a mixture with 180 angles 3. Tracker handle illumination changes 4. Robust object tracking. RMS error is 5.07 pixels per frame		

Table 2.
Challenges of pedestrian detection and tracking with related reference works.

axis may change the object appearance [29]. Tracking algorithm performance gets affected due to variation in pose. Same pedestrian looks different in consecutive frames, if the pose changes continuously. Moreover, the objects themselves may have some changes in their pose and appearance like facial expressions, changing clothes, wearing a hat, etc. Also, the target can be a nonrigid object, where its appearance may change over time. In many applications, the goal is tracking humans or pedestrians, which makes tracking algorithms vulnerable in this challenging case [30]. **Table 2** summarizes the comparative analysis of methodologies with its advantage, identified gaps and observation for handling these challenging issues in a video surveillance system.

4. Pedestrian detection and tracking

In video-based surveillance, one of the key tasks is to detect the presence of pedestrians in a video sequence, i.e., localizing all subjects that are human [45, 68]. This problem corresponds to determining regions, typically the smallest rectangular bounding boxes in the video sequence that enclose humans. In most of the surveillance systems, human behavior has been recognized using analysis of the trajectories, positions of persons and historical or prior knowledge about the scene. **Figure 5** shows some examples of pedestrian detection and tracking. Haritaoglu et al. [46] describe a combined approach of shape analysis and body tracking, and model different appearances of a person. This has been designed for outdoor environment using a single camera. The system detects and tracks groups of people and monitors the behaviors, even in the presence of partial occlusion. However, the performance is mainly based on the detected trajectories of the concerned objects in video. Furthermore, the results are not sufficient for semantic recognition of dynamic human activities and event analysis in some cases. The advanced automatic video surveillance system consists of many features such as, motion detection [69, 70],

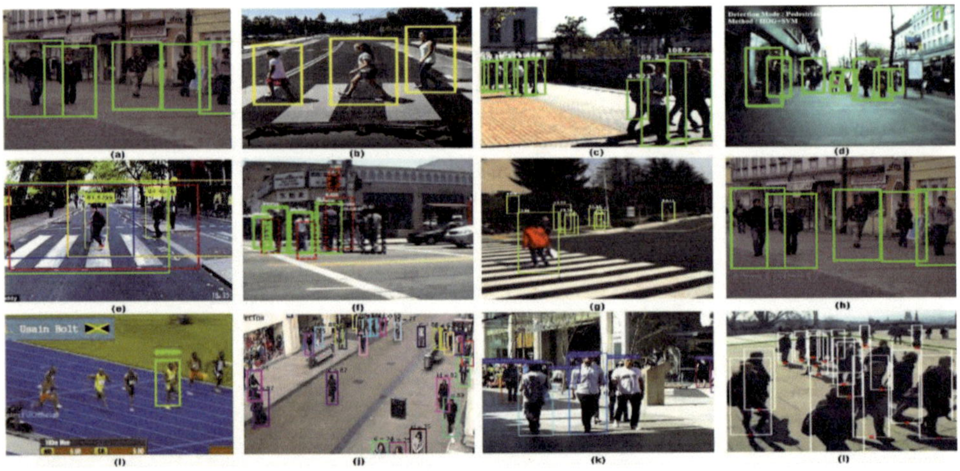

Figure 5.
Example of pedestrians detection and tracking. (a) Detecting pedestrians outdoors, walking along the street. (b) ADAS pedestrian detection. (c) Pedestrian detector is based on the aggregate channel feature detector. (d) Real-time vehicle and pedestrian detection of road scenes. (e) Pedestrian action prediction is based on the analysis of human postures in context of traffic. (f) Pedestrian detection based on hierarchical co-occurrence model (g) Cross-modal deep representations for robust pedestrian detection. (h) Pedestrian detection OpenCV. (i) Object tracking with dlib C++ library. (j) Multiple object tracking with Kalman tracker. (k) Multi-Class Multi-Object Tracking using Changing Point Detection. (l) Pedestrian tracking using Deep-Occlusion Reasoning method.

human behavior analysis, detection and tracking [71–73]. Human tracking is quite challenging, since humans may vary in intra-class variability in shape, appearance due to different viewing perspectives and other visual properties.

Krahnstoever et al. [75] designed a real-time control system of active cameras for a multiple-camera surveillance system. Hence, various researchers shifted focus from static fixed camera-based pedestrian detection to moving dynamic multi-camera-based pedestrian detection. Pedestrian tracking has been done by stationary cameras using a shape-based method [76], which detects and compares the human-body shape in consecutive frames. The cameras have been calibrated using a common site-wide metric coordinate system described in [77, 78]. Funahasahi et al. [73] developed a system for tracking the human head and face parts by means of a hierarchical tracking method using a stationary camera and a PTZ camera. The recent surveillance system focuses on human tracking by detection as described in [72–75]. Andriluka et al. [76–78] combined the initial estimate of the human pose across frames in a tracking-by-detection framework. Sapp et al. [79] coupled locations of body joints within and across frames from an ensemble of tractable sub-models. Wu and Nevatia [80] proposed an approach for detection and tracking of partially occluded people using an assembly of body parts.

The tracking of humans becomes more challenging under moving cameras than in static cameras as discussed in Section 2. Many effective pedestrian tracking techniques used in static camera, such as background subtraction and modeling [80] and a constant ground plane assumption, makes the task more difficult. Instead of using background modeling-based methods to extract the human information, human detectors are widely used to detect the human in the video. Therefore, the challenge is to successfully detect the humans in moving cameras, and then apply the tracking techniques to detected humans. However, human detectors may effectively extract human, still have some limitations viz. human detectors may produce false or miss human detection, when humans are partially or fully occluded, the detections can fail and the tracking can be unreliable until the human reappear in the frames. It is observed that, many of the researcher works on many of challenges of pedestrian detection and tracking, but still complete and reliable solution to all the challenges like discussed. Most of the algorithms of pedestrian detection and tracking were tested in indoor and outdoor environment. Attempts were also made to estimate the accuracy of the system based on detection rate, time and computational complexity. From the performance evaluation of algorithms presented in authors, it is observed that, deep learning based pedestrian detection and tracking approaches can be efficient choice for real-time environment [45, 65]. There is still a scope of improvement in existing approaches of pedestrian detection and its tracking in surveillance system.

5. Conclusions

This chapter describes and reviews the methodologies, strategies and steps involved in video surveillance. It also addresses the challenges, issue, available databases, available solutions and research trends for human detection and tracking in video surveillance system. Based on the literature survey, most of the available techniques proposed by the earlier researchers can perform object detection and tracking either within single camera view or across multiple cameras. However, most of them failed to encounter trade-off problem between accuracy and speed. Although the accuracy of the trackers is very good, they are often impractical because of their high computational requirements and vice versa. Thus, to achieve an optimal trade-off, adaptive object detection and tracking method, it is essential to achieve a real-time and reliable surveillance system. It is due to this reason that the main aim of this paper is to provide a valuable insight into the related areas of the related research topic in video surveillance and to promote new research.

Author details

Ujwalla Gawande[1*], Kamal Hajari[1] and Yogesh Golhar[2]

1 Department of Information Technology, Yeshwantrao Chavan College of Engineering, Nagpur, Maharashtra, India

2 Department of Computer Science and Engineering, GH Raisoni Institute of Engineering and Technology, Nagpur, Maharashtra, India

*Address all correspondence to: ujwallgawande@yahoo.co.in

References

[1] Minsky M, Kurzweil R, Mann S. IEEE International Symposium on Technology and Society, Toronto, Ontario, Canada, 27th–29th June; The Society of Intelligent Veillance. 2013. pp. 13-17

[2] Foresti GL, Micheloni C, Snidaro L, Remagnino P, Ellis T. Active video-based surveillance system: The low-level image and video processing techniques needed for implementation. In: IEEE Signal Processing Magazine, Vol. 22(2). March 2005. pp. 25-37

[3] Gawande U, Golhar Y. Biometric security system: A rigorous review of unimodal and multimodal biometrics techniques. International Journal of Biometrics (IJBM). April 2018;**10**(2)

[4] Cornett B. Intro to Surveillance Camera Technologies. Available at: http://www.ezwatch.com

[5] Alexandr L. IP Video Surveillance. An Essential Guide. Alexandr Lytkin. ISBN 978-5-600-00033-9. 4th April 2012

[6] Zafeiriou S, Zhang C, Zhang Z. A survey on face detection in the wild: Past, present and future. International Journal of Computer Vision Image Understand. Sep. 2015;**138**:1-24

[7] Teddy K, Lin W. A survey on behavior analysis in video surveillance applications. Video Surveillance. 2011. pp. 281-291. Available at: http://www.intechopen.com/books/videosurveillance

[8] Gawande U, Golhar Y, Hajari K. Biometric-based security system: Issues and challenges, intelligent techniques in signal processing for multimedia security. In: Studies in Computational Intelligence. Vol. 660. Cham: Springer; 2017. pp. 151-176

[9] Gawande U, Zaveri M, Kapur A. A novel algorithm for feature level fusion using SVM classifier for multibiometrics-based person identification. Applied Computational Intelligence and Soft Computing. July 2013;**2013**:1-11

[10] Dollar P, Wojek C, Schiele B, Perona P. Pedestrian detection: An evaluation of the state of the art. IEEE Transactions on Pattern Analysis and Machine Intelligence. April 2012;**34**(4): 743-761

[11] Papageorgiou C, Poggio T. A trainable system for object detection. International Journal of Computer Vision. June 2000;**38**(1):15-33

[12] MIT Pedestrian Dataset. Center for Biological and Computational Learning at MIT and MIT. 2005. Available at: http://cbcl.mit.edu/software-datasets/PedestrianData.html [Accessed: 22 September 2018]

[13] Levi D, Silberstei S. Tracking and motion cues for rear-view pedestrian detection. In: 18th IEEE Intelligent Transportation Systems Conference, Spain, 15th–16th Sept. 2015. pp. 664-671

[14] Li X, Flohr F, Yang Y, Xiong H, Braun M, Pan S. A new benchmark for vision-based cyclist detection. In: IEEE Intelligent Vehicles Symposium, Sweden, 19th–22nd. June 2016. pp. 1028-1033

[15] Campbell D, Petersson L. GOGMA: Globally-optimal Gaussian mixture alignment. In: IEEE Conference on Computer Vision and Pattern Recognition (CVPR), Las Vegas, USA, IEEE. 2016

[16] Pellegrini S, Ess A, Van Gool L. Wrong turn–No dead end: A stochastic pedestrian motion model. International

Workshop on Socially Intelligent Surveillance and Monitoring (SISM'10), (CVPR), San Francisco, CA, USA, 13th–18th June. 2010

[17] Christian W, Stefan W, Schiele B. Multi-cue onboard pedestrian detection. In: IEEE Conference on Computer Vision and Pattern Recognition, Miami, Florida, USA, 20–25 June. 2009

[18] Navneet D. Finding people in images and videos [PhD thesis]. Inria Grenoble-Rhône-Alpes; 2006

[19] Everingham M, Van Gool L, Winn WC, Zisserman. The pascal visual object classes challenge: A retrospective. International Journal of Computer Vision. June 2015;**111**(1):98-136

[20] CVC-ADAS Pedestrian dataset. 2012. Available at: http://adas.cvc.uab.es/site/ [Accessed: 22 September 2018]

[21] Lin T-Y, Maire M, Belongie S, Hays J, Perona P, Ramanan D, et al. Microsoft coco: Common objects in context. In: European Conference on Computer Vision. Springer; 2014, 2015. pp. 740-755

[22] Yun K, Lim J, Choi JY. Scene conditional background update for moving object detection in a moving camera. Pattern Recognition Letters. 2017;**88**:57-63

[23] St-Charles PL, Bilodeau GA, Bergevin R. Subsense: A universal change detection method with local adaptive sensitivity. IEEE Transactions on Image Processing. 2015;**24**(1):359-373

[24] Cogun F, Cetin AE. Object tracking under illumination variations using 2D spectrum characteristics of the target. IEEE International Workshop on Multimedia Signal Processing. 2010: 521-526

[25] Heikkila M, Pietikainen M. A texture-based method for modelling the background and detecting moving objects. IEEE Transactions on Pattern Analysis and Machine Intelligence. 2006;**28**(4):657-662

[26] Shen C, Lin X, Shi Y. Moving object tracking under varying illumination conditions. Pattern Recognition Letters. 2006;**27**(14):1632-1643

[27] Lee YB. A real-time color-based object tracking robust to irregular illumination variations. In: IEEE International Conference on Robotics and Automation, 21st–26th. May 2001. pp. 1659-1664

[28] Tokmakov P, Alahari K, Schmid C. Learning motion patterns in videos. In: IEEE Conference on Computer Vision and Pattern Recognition, CVPR. 2017. pp. 531-539

[29] Balan A, Black MJ. An adaptive appearance model approach for model based articulated object tracking. In: IEEE Conference on Computer Vision and Pattern Recognition. June 2006. pp. 758-765

[30] Porikli F, Tuzel O. Covariance tracking using model update based on lie algebra. In: IEEE Computer Society Conference on Computer Vision and Pattern Recognition, CVPR' 1. June 2006. pp. 728-735

[31] Lim J, Ross DA, Lin RS, Yang MH. Incremental learning for visual tracking. Advances in Neural Information Processing Systems. 2004:793-800

[32] Kwo J, Lee KM. Tracking of abrupt motion using Wang-Landau Monte Carlo estimation. In: European Conference on Computer Vision. Oct. 2008. pp. 387-400

[33] Zhou X, Lu Y, Lu J, Zhou J. Abrupt motion tracking via intensively adaptive Markov-chain Monte Carlo sampling. IEEE Transactions on Image Processing. 2012;**21**(2):789-801

[34] Wang F, Lu M. Hamiltonian Monte Carlo estimator for abrupt motion tracking. In: International Conference on Pattern Recognition, ICPR. Nov. 2012. pp. 3066-3069

[35] Zhang H, Zhang J, Wu Q, Qian X, Zhou T, Hengcheng FU. Extended kernel correlation filter for abrupt motion tracking. KSII Transactions on Internet and Information Systems. 2017; **11**(9):4438-4460

[36] Jepson AD, Fleet DJ, El-Maraghi TF. Robust online appearance models for visual tracking. IEEE Transactions on Pattern Analysis and Machine Intelligence. 2003;**25**(10):1296-1311

[37] Yilmaz A, Li X, Shah M. Contour-based object tracking with occlusion handling in video acquired using mobile cameras. IEEE Transactions on Pattern Analysis and Machine Intelligence. 2004;**26**(11):1531-1536

[38] Senior A, Hampapur A, Tian YL, Brown L, Pankanti S, Bolle R. Appearance models for occlusion handling. Image and Vision Computing. 2006;**24**(11):1233-1243

[39] Pan J, Hu B. Robust occlusion handling in object tracking. In: IEEE Conference on Computer Vision and Pattern Recognition. 2007. pp. 1-8

[40] Hou L, Wan W, Lee KH, Hwang JN, Okopal G, Pitton J. Robust human tracking based on DPM constrained multiple-kernel from a moving camera. Journal of Signal Processing Systems. 2017;**86**(1):27-39

[41] Delagnes P, Benois J, Barba D. Active contours approach to object tracking in image sequences with complex background. Pattern Recognition Letters. 1995;**16**(2):171-178

[42] Monnet A, Mittal A, Paragios N, Ramesh V. Background modelling and subtraction of dynamic scenes. In: Ninth IEEE International Conference on Computer Vision. 2003. pp. 1305-1312

[43] Li L, Huang W, Gu IY, Tian Q. Statistical modelling of complex backgrounds for foreground object detection. IEEE Transactions on Image Processing. 2004;**13**(11):1459-1472

[44] Chetverikov D, Péteri R. A brief survey of dynamic texture description and recognition. Computer Recognition Systems. 2005:17-26

[45] Arashloo SR, Amirani MC, Noroozi A. Dynamic texture representation using a deep multi-scale convolutional network. Journal of Visual Communication and Image Representation. 2017;**43**:89-97

[46] Minematsu T, Uchiyama H, Shimada A, Nagahara H, Taniguchi RI. Adaptive background model registration for moving cameras. Pattern Recognition Letters. 2017:86-95

[47] Xia H, Shuxiang S, Liping H. A modified Gaussian mixture background model via spatiotemporal distribution with shadow detection. Signal, Image and Video Processing. 2016;**10**(2): 343-350

[48] Song R, Liu M. A shadow elimination algorithm based on HSV spatial feature and texture feature. In: International Conference on Emerging Internetworking, Data Web Technologies. 2017. pp. 585-591

[49] Treptow A, Zell A. Real-time object tracking for soccer-robots without color information. Robotics and Autonomous Systems. 2004;**48**(1):41-48

[50] Hua C. A noise-insensitive object tracking algorithm. In: Conference on Computer Vision. Nov. 2007

[51] Unger M, Asdsch M, Hosten P. Enhanced background subtraction using global motion compensation and

mosaicing. In: International Conference on Image Processing, ICIP. Oct. 2008. pp. 2708-2711

[52] Li Y, Ai H, Yamashita T. Tracking in low frame rate video: A cascade particle filter with discriminative observers of different life spans. IEEE Transactions on Pattern Analysis and Machine Intelligence. 2008;**30**(10):1728-1740

[53] Spampinato C, Chen-Burger YH, Nadarajan G, Fisher RB. Detecting, tracking and counting fish in low quality unconstrained underwater videos. VISAPP. 2008;**2**:514-519

[54] Wu Y, Ling H, Yu J, Li F, Mei X, Cheng E. Blurred target tracking by blur-driven tracker. In: International Conference on Computer Vision. Nov. 2011. pp. 1100-1107

[55] Dollár P, Appel R, Blondie S, Perona P. Fast feature pyramids for object detection. IEEE Transactions on Pattern Analysis and Machine Intelligence. 2014;**36**(8):1532-1545

[56] Zhang X, Hu W, Xie N, Bao H, Maybank S. A robust tracking system for low frame rate video. International Journal of Computer Vision. 2015;**115**(3):279-304

[57] Chen HK, Zhao XG. heterogeneous features fusion-based low-resolution human detection method for outdoor video surveillance. International Journal of Automation and Computing. 2017;**14**(2):136-146

[58] Irani M, Anandan P. A unified approach to moving object detection in 2d and 3d scenes. IEEE Transactions on Pattern Analysis and Machine Intelligence. 1998;**20**(6):577-589

[59] Sawhney HS, Guo Y, Kumar R. Independent motion detection in 3d scenes. IEEE Transactions on Pattern Analysis and Machine Intelligence. 2000;**22**(10):1191-1199

[60] Zhou D, Frémont V, Quost B, Dai Y, Li H. Moving object detection and segmentation in urban environments from a moving platform. Image and Vision Computing. 2017;**68**:76-87

[61] Available from: http://www.cvpapers.com/datasets.html

[62] Wang J. Representing moving images with layers. IEEE Transactions on Image Processing. 1994;**3**(5):625-638

[63] Xiao J, Shah M. Motion layer extraction in the presence of occlusion using graph cut. IEEE Transactions on Pattern Analysis and Machine Intelligence. 2005;**27**(10):1644-1659

[64] Chen T, Lu S. Object-level motion detection from moving cameras. IEEE Transactions on Circuits and Systems for Video Technology. 2017;**27**(11):2333-2343

[65] Hayman E, Eklundh JO. Statistical background subtraction for a mobile observer. In: IEEE International Conference on Computer Vision, ICCV. 2003. pp. 67-74

[66] Shen Y. Video stabilization using principal component analysis and scale invariant feature transform in particle filter framework. IEEE Transactions on Consumer Electronics. 2009;**55**(3):1714-1721

[67] Li-Fen T, Qi P, Si-Dong Z. A moving object detection method adapted to camera jittering. Journal of Electronics and Information Technology. 2014;**35**(8):1914-1920

[68] Black MJ, Jepson AD. Eigentracking: Robust matching and tracking of articulated objects using a view-based representation. International Journal of Computer Vision. 1998;**26**(1):63-84

[69] Athanesious SP. Systematic survey on object tracking methods in video. International Journal of Advanced

Research in Computer Engineering and Technology. 2012;**1**(8):242-247

[70] Chaaraoui A, Climent-Perez P. A review on vision techniques applied to human behavior analysis for ambient-assisted living. Expert System Application. Sep. 2012;**39**(12): 10873-10888

[71] Hu W, Tan T, Wang L, Maybank S. A survey on visual surveillance of object motion and behaviors. IEEE Transaction on Systems, Man, and Cybernetics, Part C. 2004;**34**:334-352

[72] Haritaoglu I, Harwood D, Davis L. Real-time surveillance of people and their activities. IEEE Transaction on Pattern Analysis and Machine Intelligence. 2000;**22**:809-830

[73] Hu W, Tan T, Wang L. A survey on visual surveillance of object motion and behaviors. IEEE Transaction on Systems, Man, and Cybernetics, Part C. Aug. 2004;**34**(3):334-352

[74] Tsai D-M, Lai S-C. Independent component analysis-based background subtraction for indoor surveillance. IEEE Transaction on Image Processing. Jan. 2009;**18**(1):158-167

[75] Stauffer C, Grimson WEL. Learning patterns of activity using real-time tracking. IEEE Transaction on Pattern Analysis and Machine Intelligence. Aug. 2000;**22**(8):747-757

[76] Yilmaz A. Object tracking: A survey. ACM Computing Survey. Dec. 2006; **38**(4):1-45

[77] Wu D, Shao L. Silhouette analysis-based action recognition via exploiting human poses. IEEE Transaction on Circuits Systems and Video Technology. Feb. 2013;**23**(2):236-243

[78] Li L, Huang W, Gu IY-H, Tian Q. Statistical modeling of complex backgrounds for foreground object detection. IEEE Transaction on Image Processing. Nov. 2004;**13**(11):1459-1472

[79] Krahnstoever N, Yu T, Lim S-N, Patwardhan K, Tu P. Collaborative real-time control of active cameras in large scale surveillance systems. In: Proceeding Workshop, France. Oct. 2008. pp. 1-12

[80] Elaiw A, Al-Turki Y, Alghamdi M. A critical analysis of behavioural crowd dynamics—From a modelling strategy to kinetic theory methods. MDPI Symmetry Journal. July 2019;**11**(851): 1-11